T0264961

Loudspeaker Modelling and Design

In this book, Geoff Hill demonstrates modern software and hardware being applied to the processes behind loudspeaker design and modelling. Modern computing power has progressed to the point that such analyses are now practical for any interested individual or small company. *Loudspeaker Modelling and Design: A Practical Introduction* examines the process from initial concept through specifications and theoretical simulations and onto detailed design. It demonstrates the processes of design and specification, by using detailed simulations of a loudspeaker driver; sufficient to give re-assurance that a design is practical and will perform as expected. This book brings together many different strands of modelling from electro-magnetic through to mechanical and acoustic, without getting bogged down in detailed theoretical discussions and arguments. This practice-based book shows the techniques used in designing modern loudspeakers and transducers.

Geoff Hill is a passionate engineer, active member of the Audio Engineering Society, and a consultant working with a wide range of companies in the UK, Europe, and Asia. His latest venture, Hill Acoustics, marks in many ways a return to his roots, combining his 40+ years of experience with a determination to make a difference and do things better. Geoff can be contacted at www.geoff-hill.com or through his business www.hillacoustics.com.

Loudspeaker Modelling and Design

A Practical Introduction

Geoff Hill

Routledge
Taylor & Francis Group

NEW YORK AND LONDON

First published 2019
by Routledge
711 Third Avenue, New York, NY 10017

and by Routledge
2 Park Square, Milton Park, Abingdon, Oxon, OX14 4RN

Routledge is an imprint of the Taylor & Francis Group, an informa business

© 2019 Taylor & Francis

The right of Geoff Hill to be identified as author of this work has been asserted by him in accordance with sections 77 and 78 of the Copyright, Designs and Patents Act 1988.

All rights reserved. No part of this book may be reprinted or reproduced or utilised in any form or by any electronic, mechanical, or other means, now known or hereafter invented, including photocopying and recording, or in any information storage or retrieval system, without permission in writing from the publishers.

Trademark notice: Product or corporate names may be trademarks or registered trademarks, and are used only for identification and explanation without intent to infringe.

First edition published by the author 2015

Library of Congress Cataloging-in-Publication Data
Names: Hill, Geoff, author.
Title: Loudspeaker modelling and design : a practical introduction/Geoff Hill.
Description: New York, NY: Routledge, [2019] | Includes bibliographical references.
Identifiers: LCCN 2018020385| ISBN 9780815361329 (hardback : alk. paper) | ISBN 9780815361336 (pbk. : alk. paper) | ISBN 9781351116428 (ebook)
Subjects: LCSH: Loudspeakers–Design and construction.
Classification: LCC TK5983 .H55 2019 |
DDC 621.382/84–dc23LC record available at https://lccn.loc.gov/2018020385

ISBN: 978-0-8153-6132-9 (hbk)
ISBN: 978-0-8153-6133-6 (pbk)
ISBN: 978-1-351-11642-8 (ebk)

Typeset in Minion Pro
by Sunrise Setting Ltd, Brixham, UK

Dedication

To Su and Lilly for all your love and support
To all at the London Heart Hospital who gave me the opportunity to write this book
To Pete Wright for help and assistance with LATEX
To Paul Messenger for his skill and patience in editing
To Lyn Nesbitt-Smith for her skill and patience in indexing

To all of the authors and programmers whose work this book is directly built,
including:
Joerg Panzer for ABEC and VACS
Autodesk for Fusion 360
David Meeker for FEMM
Thomas Holm for HOLMImpulse
Wolfgang Klippel for Klippel
Victor Kemp for Mecway
Spectrum Software for Micro-Cap
Patrick Macey for PAFEC
Juha Hartikainen for WinISD
Thanks for the encouragement, feedback and reviews

To Alan, Dave, Gary, Kelvin, Steve and Ivo for further comments and reviews
All mistakes are mine alone

To everyone and the many companies I have worked with over the years, especially
to Wolfgang Klippel for permission to include many examples.
And most importantly to the reader: may you gain some of the joy I have had
working in this field!

Contents

Figures

Tables

PART I

Introduction

Preface to Second Edition

This second edition concentrates more deeply on the core of modelling loudspeaker transducers. To do this I have relaxed the use of open source software, including a mix of commercial products.

Included are Klippel, CLIO, PAFEC, and the Tetrahedral Test Chambers.

Klippel's dB-Lab and Analyser were of course a crucial part of explaining many of the key concepts in the first edition. This second edition goes into more detail, with a tutorial that demonstrates dB-Lab alongside a small-signal parameter analysis using the LPM module.

Audiomatica's CLIO Pocket shows what can be achieved with more modest hardware and software. I show examples of SPL versus frequency response and with respect to time, as this shows the decay behaviour.

Pafec appears via the PafLS front end and undertakes loudspeaker driver simulation of impedance as well as SPL at selectable angles, showing how a loudspeaker drive unit behaves off-axis.

I introduce Autodesk's Fusion360, which directly builds a 3D model, produces drawings (including 2D examples), and conducts mechanical simulations.

The entire manuscript has had the benefit of Paul Messenger's editorial eye for detail and years of writing, and I am deeply grateful for his time and work here. Nevertheless, any remaining mistakes are mine and mine alone. I am also deeply grateful to Lyn Nesbitt-Smith for a comprehensive index.

There is a greatly expanded glossary of terms, and a list of references after each main chapter, where these refer to individual books, articles, or patents, etc.

Most of the simulations and models have been revised, and many of the 3D models have been redrawn using Fusion360. An additional chapter, Linkwitz Transform, rounds off the subwoofer design.

This second edition greatly benefits from the care and attention of the entire team, who will be able to bring this to a wider audience.

One thing I would stress is that, as an engineer, things always change. How something was done a few years or decades ago will be different; however, the fundamentals remain constant.

Although the core remains the same, I believe there is a gap between theoretical knowledge as it is often taught and real engineering where such theoretical knowledge is translated into actual products.

This is especially true now, when so many loudspeakers use 'off the shelf' loudspeaker drive units.

One of the things I do is try to bridge that gap, and I am interested in helping others do so too.

So while there still is some theory in this book, I have tried to ensure you do not get bogged down in it: neither do you need to be an expert in computer modelling, mathematics, or measurements.

I hope the tutorials are self explanatory and complete in themselves. If you don't understand the detailed theory or modelling, don't worry: just pick and choose the bits you need at the time.

I would particularly like to thank readers of the first edition who have written directly to me; I have tried to incorporate your feedback in this second edition. Necessarily, this work is ongoing, and some of the comments will have to wait for subsequent editions.

I should like to acknowledge the support and assistance of all the authors and programmers of the software and hardware used in this edition.

Lastly I should like to repeat my request for feedback on this edition. You can contact me directly at geoff@geoff-hill.com.

Geoff Hill August 2018

Aim of This Book

0.1 Aim of This Book

The main aim of this book is to bring together the disparate strands that are required to design and specify current loudspeaker drive units, all in one place. We believe that this book is long overdue, so if you don't want to study hundreds of papers and pick out the relevant nuggets, this book may well be for you.

The book brings together nearly 40 years' experience of actually designing loudspeaker drivers and systems. It's fully up to date and is backed up by the development of both systems and drive units. Our aim is to concentrate mainly on the practical aspects of loudspeaker design, simulation, and measurement, rather than the theory.

There is a little theory, but we have tried to keep this at a fairly basic algebraic level, by leaving all the complicated calculations to the individual programs themselves. A key idea is to take the reader through the whole process required to design and specify loudspeaker drivers and ultimately speaker systems.

The performance of any finished loudspeaker is dominated by the qualities or problems of the underlying driver(s).[1] So the specification and simulation of the drivers need to be understood before we can design and produce high quality loudspeakers.

Loudspeaker drivers are currently used in the following devices: ear-buds, headphones, computers/tablets, telephones and mobiles (including smartphones), radios, TVs, cinemas, hi-fi and home cinema systems, musical instruments, and public address systems for concerts and festivals.

All of these need one or more of the following units: micro speakers, full range drivers, tweeters, midrange and bass units, subwoofers, and compression drivers. It's quite a list, and society today would be almost unrecognisable without the capability to reproduce sound and music.

The fact is that many loudspeakers have been around in their current form for nearly 100 years. The vast majority are based upon the moving-coil design, and most (but not all) can trace their roots back to the Rice & Kellogg [5] patents issued between 1924 and 1929.

This book will take the reader through the design process as we see it, building from the initial concept to its specification(s) and the theoretical modelling of the transducer itself. It will then continue to the detailed design, test, and measurements, plus the statistical analysis of the final product.

We will do this using open-source, free, and commercial software using modern computing techniques. Only in the past 20 years have the modern analytical tools and computing power become readily available, and it is only in the last decade that these tools have started to trickle down to the point that an individual can have access to both the knowledge and the software.

This step change in the availability of knowledge and capability is unprecedented, and it is the subject of progress in many fields in computing and software. Think of the changes Linux has wrought in mobile phones, through the Android and Apple MAC operating systems to 3D CAD and printing.

The target audience includes those people who are just interested in loudspeaker driver design as well as students and engineers who design the products we use on a daily basis.

Much has been written about the theoretical aspects of loudspeaker design, some of which has been published in journals such as the Audio Engineering Society, the Acoustical Society of America, the Institute of Acoustics, and others. There are also many excellent reference books on acoustic theory, from Rayleigh [4], Morse [3], Beranek [1], Kinsler & Frey [2], and many others. Unfortunately, most of this has been from a purely theoretical perspective, which has put off many people from further exploration because of the 'academic' approach this implies.

Sometimes it may be necessary to develop a loudspeaker fully from theory, but in practice much real design is based upon tried and tested developments, often made by relatively inexperienced engineers thrown into a task at short notice. How to approach the tasks then?

In reality, a lot of the real design work is (and perhaps always has been) done by relatively inexperienced engineers or technicians perhaps a year or two out of college or university. They may well have been taught the academic theory by their lecturers, but do they really need to start from scratch? We would say no. That is a very poor choice today and leaves many points unknown, simply because those who are teaching design have (often) never actually done it! Another case is where designers or engineers are experienced, but in a different area, and they have a need to understand how to design and model loudspeakers.

Much has been written about fitting or matching drivers to cabinets as well as crossover design and system integration, and this information tends to be relatively accessible.

We will therefore mainly limit our explanations to those required to understand techniques relevant to designing loudspeakers and drivers. We make no apologies if these explanations are non-standard. We find them particularly useful in understanding various techniques (rather more so than theoretically exact explanations).

We think and hope that there may be some demand for this approach. We will try to give specific worked examples by actually using various product design tools, a few of which date back 25 or more years. (We should also point out that this book may not be for those who are already fully conversant with the art of loudspeaker design.) We do this to demonstrate the *practice* of loudspeaker design and measurement.

Even so, we hope that it will demonstrate one or two hardware/software techniques that even those with a reasonable knowledge of loudspeaker design were not aware of. Indeed, many of the reviewers have commented how useful many of these techniques were to them personally.

We will try to be open and honest in our approach, especially when discussing potential alternatives. As so often, it is balancing one thing against another to best effect where the art of loudspeaker design gets most interesting.

As much of this book will involve using equipment or software for these tasks, some people will doubtless point out that it is the job of software and hardware suppliers to provide the instructions for their products. Here we both agree and disagree. If we were talking about software and products designed and sold for a particular task, we would tend to agree. But with open-source, free, or low-cost software this is neither possible nor practical, so we will try to fill the gap. Even with the commercial equipment, the focus tends to be on the unusual or difficult tasks rather than the essential everyday ones.

In any case, the multi-disciplinary nature of loudspeaker design means that it is not the job of any of the 'tool makers' to explain the process of speaker design itself. There's an argument that colleges and universities should do so, but speaker design is only a small part of a degree course in acoustics (let alone one in physics).

One realistic result of these pressures is that most research is undertaken and published by individual companies. However, since these companies inevitably need to maintain a competitive advantage, it can be argued that it is not in their interest to reveal how and why they do essential loudspeaker design.

The obvious exception is where companies publish a patent, whereby they are granted exclusive right to use the patented idea for a period in return for making the information publicly available. But even here there is a hidden agenda in the need to keep as much other information as possible secret to hinder potential competition.

This conflict of interest (which happens to a greater or lesser extent in all industries) may well explain why people today seem to be disengaged from the process of designing and making things.

We fully accept that most people are not interested in how things work or how to design things, but as you are reading this, it will probably not apply to you. However, it is difficult for someone new to loudspeaker design to get started, so this book attempts to demonstrate some of the techniques currently used in modern loudspeaker design.

We should state that although we have taken great care preparing this book and its contents, neither we nor the publishers can accept any responsibility for errors, or any losses or damage, however caused. The risk is and will remain yours alone.

However, we would welcome feedback on any errors or omissions you find, so that we may correct and update these in the future. Our aim is to concentrate mainly on practical loudspeaker design, simulation, and measurement, rather than the theory of moving-coil loudspeakers.

0.2 What This Book Is Not

This book is not intended to replace or supplant in any way the theory or academic methods taught in schools, colleges, or universities. Neither does it look at other types of loudspeaker driver designs such as moving magnet, piezoelectric, electrostatic system, or many others, and it only touches on system integration, crossovers, and many other topics.

We will not even try to prove any of the fundamental concepts backing the various aspects of design that would be required by many such courses. We will, however, do our best to explain them, often in quite simple terms, and especially when this helps in understanding or directly designing loudspeaker drivers.

The techniques presented do all work, and many of them are demonstrated in detail in the various tutorials at the end of the book. Some techniques have been developed and refined over many years, and those using spreadsheets or other programs have been rewritten from scratch using up to-date versions of software to help their longevity. Others are newer, and we hope to incorporate them into future projects.

Neither, unfortunately, does this book guarantee that you will design the best loudspeaker drivers and speaker systems possible. It has been said that the difference between science and art is that art has more than ten variables—so there is maybe some truth in the notion that loudspeakers are at least in part a form of art.

What we will try to do is to point out some of the pitfalls on the way and demonstrate a few of the techniques that we have found useful.

Also, apart from specific tutorials for some of the software that we have used, we are not aiming to teach any particular subject in ultimate depth—partly because this would not be possible.

But more importantly, it is necessary to provide an overall background of many of the areas that impact loudspeaker design. If you need further in-depth information, the best we can do is to point you in the direction of books, courses, or online resources that we have found useful.

Please let us be clear about one other thing: we are not saying that the ways we are describing are the only, or even necessarily the best, ways of doing things. Rather, we have found these techniques helpful in designing loudspeaker drivers, and we hope they can be useful to you and others in the future.

Another and perhaps one of the most important points: this book does not aim to be perfect. There are and will be better ways to be found in the future, of that we are sure.

So we will show you our workings throughout—even when things do not work—so that you may learn and profit from these mistakes and misunderstandings. As the saying goes: 'Learn from your mistakes and not from your successes.'

Note

1. DSP can correct some problems, but these are strictly limited.

References

[1]   Leo Leroy Beranek and Tim J. Mellow. *Acoustics: sound fields and transducers*. Academic Press, 2012.
[2]   Lawrence E. Kinsler et al. *Fundamentals of acoustics*. John Wiley & Sons, 2000.
[3]   Philip McCord Morse and K. Uno Ingard. *Theoretical acoustics*. Princeton University Press, 1968.
[4]   John William Strutt Baron Rayleigh. *The theory of sound*. Vol. 2. Macmillan, 1896.
[5]   Chester W. Rice and Edward W. Kellogg. *Loud Speaker*. US Patent 1,707,570. Apr. 1929.

Tools and Techniques

This chapter will focus on the tools and techniques now available for designing loudspeakers. These have changed significantly over the past ten years, so many earlier examples are now effectively obsolete. Some, like the Thiele/Small parameters, have morphed into mainstream design and specifications, if not always in the way originally intended. This chapter also outlines the key categories of software and hardware that we consider to be minimum requirements.

The focus of this book will be on using modern and freely available tools, such as free or open-source software wherever possible, although commercial software and better known equipment take a larger part in this edition. Our aim is to make the book relevant and usable to a wide readership.

Many of the techniques are of course equally applicable to other software or hardware, be it commercial, free, or open source; using these other options may require working a different way to get the functionality needed. Hopefully the techniques and potential problems highlighted will shine through.

Please consider this book as a compendium of techniques that we have found useful and that hopefully you will too. We will focus on the implementation of the aspects required to design and produce a loudspeaker, but there are so many possibilities it would be impossible to cover them all.

The advent of cheap computing power and the internet has meant that knowledge is now more readily available than ever before. Couple this with recent developments in small-scale manufacturing and prototyping, and it becomes possible for a single person or small company to design, assemble, test, and ultimately enjoy listening to a complete loudspeaker system of potentially far higher quality than ever before. Readily available computing power has been the driver behind many modern techniques, such as:

- Digital signal processing (DSP).
- Computer aided design (CAD).
- Computer aided manufacturing (CAM).
- Software packages.
- Automatic test equipment.
- Rapid manufacturing techniques.

Let us remember that none of these methods existed when loudspeakers were first developed. Throughout this book we'll demonstrate how these tools (all readily available) can do the various tasks needed. Inevitably the work will be incomplete, as we cannot possibly cover all the possible ways of doing things. If you know of a better way to do something, or a better and more useful tool (especially free or open source), please feel free to contact us, as we are always eager to learn.

0.3 Modelling Approaches

One approach is to model everything together as a whole, which it actually is. This means taking a holistic approach in which all of the subsections are modelled in a common way with data exchanged bidirectionally between the different modelling domains. Obviously, this requires that the modelling tool used is capable of operating in all of these physical domains.

This is known as a *multiphysics fully coupled model*. Software packages are available that are capable of doing this; however, they tend to be expensive and professional systems, such as ANSYS® [2], COMSOL Multiphysics® [6], and PAFEC [23]. In theory, open-source software packages such as CAELinux [4] and Elmer [8] should also be capable of this level of modelling. PafLS is a template-driven version of PAFEC; although fully coupled, it models only some of the domains: specifically, mechanical to acoustical.

Alternatively, we could look at and analyse a loudspeaker as a machine, breaking it down into smaller subsections, each of which can be considered semi-independently. Ultimately, these subsections are connected, but we can go a long way in our understanding by considering them independently; we can then understand these parts individually and finally stitch them together to form the whole.

This book will mainly approach modelling a loudspeaker from the perspective of breaking things down into individual sections and subsections. This will enable us to deal with them on a simple level, which can be more useful than having a large model with many elements. On occasion this approach may break down, but we will deal with this when it occurs.

The fundamental loudspeaker equations will define the essential parameters, along with the low-frequency response. These in turn will give us data that we can use to define and design the proposed parts:

- We will use CAD, Magnetic FEA, and spreadsheet(s) for the motor unit design, including the magnet, top plate, back plate, and voice coil.
- We will use CAD, Mechanical FEA, and spreadsheet(s) for the mechanical to acoustical design, including the voice coil, diaphragm, dust cap, surround, and centralisation parts.
- We will use CAD, Acoustic BEM, and spreadsheet(s) for the acoustical simulation and final output(s). These will include the diaphragm, cabinet, vent, and the environment.
- We will use PafLS to model the mechanical to acoustical domain of a driver on a baffle.
- We will use amplifiers, analysers, microphones, sensors, measurement software, spreadsheets, and test chambers to confirm the performance of our designs against our initial predictions.

0.4 Professional Equipment

All of the consumer electronics, mobile phone, or loudspeaker companies that we have worked for have invested significant resources in measurement and or analysis equipment from companies such as:

- Audio Precision [3].
- CLIO (Audiomatica) [5].
- Etani [9].
- Klippel [15].
- ListenINC [16].
- Loudsoft [10].
- MLSSA [19].
- NTi Audio [20].

0.5 Software

We have mostly used Windows® programs during the course of this book, and we have kept largely to open-source, free, or at least low-cost software. Please see the internet addresses and links in section B.2 on page 187 for further details.

You may well ask why just Windows®. This is simply because we are more familiar with this operating system and have focused upon using programs and methods that work. Fortunately, many, although not all, of the more up-to-date programs are either directly available on Mac OSX® or Linux systems. (For those that are not available then there are options like Wine,[1] Crossover, Parallels, VMWare, or VirtualBox, which between them should run most of these programs on Linux and Mac OSX®; equivalent tools are now becoming available which will run on Android and iOS systems.)

Increasingly there are some like Fusion 360 [12] and OnShape [21] that are completely independent of the operating system, being cloud-based and operating with access provided by web browsers and apps.

Within this book, we have used the following programs and software at various times for different tasks. We appreciate the work and time that all of these authors have put into their respective programs, and we salute you all.

- 2D CAD programs:
  - Fusion 360 [12].
  - DraftSight ® [7].
- 3D CAD programs:
  - Fusion 360 [12].
  - OnShape [21].
- Electromagnetic Finite Element Analysis (FEMM) [11].
- Mechanical/Flexural/Modal Finite Element Analysis:
  - Fusion 360 [12].
  - Mecway [17].
- Acoustic FEA (Pistonic) & Acoustic Display (Fields): ABEC [1].
- Acoustic Display (Spectrum, Polar & Contour Graphs): VACS [25].
- Mechanical–Acoustical Modelling & Analysis: PafLS [23].
- Spreadsheet: Open Office [22] or Microsoft® Excel [18].
- Loudspeaker Design:
  - SpeakerPro [13].
  - WinISD [26].
- Measurement Tools:
  - CLIO Pocket [5].
  - HolmImpulse [14].
  - Klippel [15].
  - REW [24].

0.6 Hardware

0.6.1 Sound Card

If you do not have access to professional equipment such as shown in section 0.4 on page xxxviii then as a minimum we would recommend an external USB/Firewire (or better) sound card; this should be a good quality example, preferably with balanced floating dual inputs. Although one could use an internal sound card (or the computer itself), there are several reasons why this is not a good idea:

1. You really will get better results with an external card.
2. It is not unknown for test amplifiers to put out high voltages.
3. Many amplifiers run in BTL (bridge tied load), so there's the risk of shorting out the amplifier or computer (maybe even both).
4. Computers have lots of very high-frequency signals floating around, and these are better kept out of our nice clean signal.

We would suggest that you get one with at least two completely independent balanced inputs, ideally fully floating. You will also need at least one output. Ensure that the selected external card can run in full Duplex mode.[2]

If using phantom powered microphones, the sound card needs to provide the requisite power. The bit depth and sample rate will depend on the measurement needs: we recommend at least 96 kHz.

Other key criteria for a sound card include low distortion, low input channel crosstalk, and I/O clock synchronisation (for correlation methods).

One useful tip here: If your sound card has adjustable controls, please 'lock' them in position after you have run any calibration.[3] If you cannot lock them any other way, then please consider a dedicated card and fix the settings by glueing the controls in position. Otherwise, it is all too easy to get different results the next time you undertake a measurement.

Examples that we have successfully used include M-Audio Transit and MobilePre, and Roland UA-25 EX. Many others should be suitable as well.

0.6.2 Power Amplifier

The power amplifier needs to have the following characteristics:

- A fixed (for some software or analysers it needs to be adjustable in steps) gain.
- We would suggest building or purchasing a fixed stepped attenuator to use with most amplifiers.
- A frequency response ±0.2 dB from D.C. up to 50 kHz or ideally 100 kHz.
- It should have low noise and distortion.
- It must have sufficient power to run the tests that you intend to conduct.
- It should not make any noises turning it on or off.
- It needs to be unconditionally stable.
- It *must* be fully protected against open circuits, short circuits, and shorting to ground. This is especially important for BTL amplifiers as it is very easy to inadvertently short these out.

Further options could include selectable low and high output impedance settings, as these would enable both SPL and impedance response measurements under different driving conditions. This could be crucial if your loudspeaker is intended to be used with a high-output impedance amplifier.

0.6.3 Sound Calibrator

Most companies need to be able to define the sound pressure level (SPL) from a loudspeaker or a speaker system. Then you need a known sound level to calibrate the measurement microphone.

This is provided by a microphone level calibrator. These typically output a known level, usually 94 dB or 114 dB at 1000 Hz, which is used to provide a known sound pressure level for your test microphone.[4]

These used to be relatively expensive, but today simple third-party designs are available for around £100 or $100. Calibration to ensure international traceability will cost a bit more and will usually need to be renewed annually or biannually.

0.6.4 Microphone

What you need in a measurement microphone is very dependent upon what your requirements are. We are concerned with designing and measuring loudspeakers—so to us, focussing upon a very low noise floor in the microphone is *much* less important than the high-frequency performance and measurement capability at high sound pressure levels.

Generally, the larger a microphone's diaphragm is physically, the lower the noise floor and the higher the microphone's sensitivity. However, the high-frequency performance is often worse; also, smaller microphones tend to be able to handle higher sound pressure levels before distorting.

As even the most inefficient speaker will produce 70 dB or 80 dB SPL, concentrating on noise floors at 0 dB SPL or below makes little sense for measurement of loudspeakers, so we recommend using the smaller microphone sizes of half an inch or below.

The minimum microphone specification we would recommend would be a Behringer ECM8000 (ideally with both frequency *and* phase response correction data).

0.7 Impedance Measurements

0.7.1 High Impedance

This is also known as constant current mode.

With the high impedance method, a large, typically 1000 ohm, resistor is often connected in series with the loudspeaker, and the impedance is directly read off as a voltage. This works as:
(i) The peak impedance of the loudspeaker is typically much less than 1000 ohms, so it *effectively* becomes a straightforward voltage divider.
(ii) The 1000 ohm resistor ensures that the impedance measurement is being made in the small signal domain.
(iii) In many cases, people use the high output impedance of a 470 ohm resistor after an operational amplifier (op-amp) for this; note, however, this impedance is only rarely accurately known, which can lead to errors.

The circuit is shown in figure 0.1.

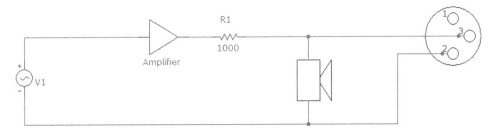

Figure 0.1: Measuring Impedance Using High Impedance Method.

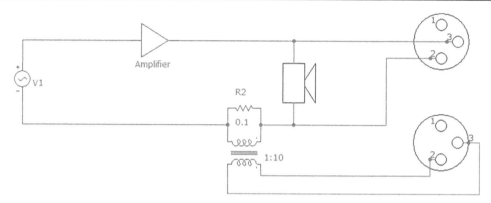

Figure 0.2: Measuring Impedance Using Low Impedance Method.

0.7.2 Low Impedance

The low impedance method is also known as constant voltage mode. Here we minimise the series impedance in the output circuit, ensuring that the loudspeaker is voltage controlled. This method enables measurements to be made under high power conditions. Here we use a 0.1 ohm or 0.01 ohm ground resistor, or ideally a *zero* impedance current sensor, thereby directly monitoring the current through the loudspeaker under test.

A quick division of voltage (across the loudspeaker) by the current flowing through the speaker will, of course, reveal the impedance very easily in modern software. Back in 1990 or so we wrote BASSCALC; this made automatic parameter calculations of Thiele/Small parameters for the Audio Precision System One using a 1:10 microphone transformer both to isolate the ground return signal and to boost the signal level so that a straight division could be made.

The circuit is shown in figure 0.2.

Please note the use of 'balanced' and floating input pins 2 and 3 on the 3-pin XLRs. These will ensure that nothing gets shorted out—a real problem with direct connections into any computer/tablet.

0.7.3 Known Impedance

Lastly, there is the known impedance method, also known as the voltage ratio method. Here we measure the voltage at both ends of a *known* reference resistor (R), the unknown impedance being called (Z). This is shown in figure 0.3.

The voltage ratio method can have one *major* disadvantage in that it is possible to use this with any known resistor, thus inadvertently changing the damping ratio.

However, this can be used to your advantage if you need to measure how a loudspeaker would work with a valve amplifier.

$$\text{Unknown Impedance (Z)} = \text{Voltage Out} * R/(\text{Voltage In} - \text{Voltage Out}) \qquad (0.1)$$

Between them, these three methods enable you to calculate the impedance and later the parameters at low or high signal levels. There can be a considerable difference, as we shall see when we get to later chapters on large signal parameters.

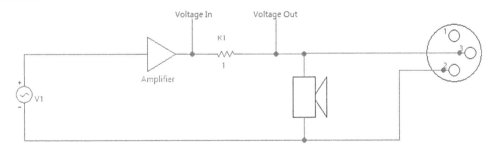

Figure 0.3: Measuring Impedance Using Known Impedance Method.

Notes

1. Wine is available from www.winehq.org.
2. Duplex mode means that it is capable of working both inputs and outputs simultaneously.
3. Unfortunately, most 'high quality' sound cards do not offer this option.
4. Although 114 dB at 250 Hz used to be common.

References

[1]   *ABEC.* www.randteam.de/ABEC3/Index.html (visited on 02/02/2018).
[2]   *ANSYS Engineering Simulation.* www.ansys.com/ (visited on 31/01/2018).
[3]   *Audio Precision.* www.ap.com/ (visited on 28/02/2018).
[4]   *CAE Linux.* http://caelinux.com/CMS/ (visited on 31/01/2018).
[5]   *CLIO Pocket.* www.audiomatica.com/wp/?page_id=1739 (visited on 01/02/2018).
[6]   *COMSOL Multiphysics.* www.comsol.com/ (visited on 31/01/2018).
[7]   *Draftsight 2D CAD.* www.3ds.com/products-services/draftsight-cad-software/free-download/ (visited on 01/02/2018).
[8]   *ELMER.* www.csc.fi/web/elmer (visited on 31/01/2018).
[9]   *ETANI.* www.etani.co.jp/eng/ (visited on 28/02/2018).
[10] *Fine Speaker Test & Design.* www.loudsoft.com/ (visited on 01/03/2018).
[11] *Finite Element Method Magnetics.* www.femm.info/wiki/HomePage (visited on 31/01/2018).
[12] *Fusion 360 ™ 3D CAD, CAM, and CAE.* www.autodesk.com/products/fusion-360/free-trial (visited on 31/01/2018).
[13] *Geoff Hill website.* www.geoff-hill.com/ (visited on 02/02/2018).
[14] *HOLM Impulse.* www.holmacoustics.com/holmimpulse.php (visited on 31/01/2018).
[15] *Klippel.* www.klippel.de/ (visited on 31/01/2018).
[16] *ListenInc.* www.listeninc.com/ (visited on 28/02/2018).
[17] *Mecway Finite Element Analysis for Windows.* http://mecway.com/ (visited on 31/01/2018).
[18] *Microsoft Office Excel.* https://products.office.com/en-GB/business/get-office-365-for-your-business-with-latest (visited on 31/01/2018).
[19] *MLSSA Acoustical Measurement System.* www.mlssa.com/ (visited on 04/02/2018).
[20] *NTi Audio.* www.nti-audio.com/en/home.aspx (visited on 28/02/2018).
[21] *OnShape.* www.onshape.com (visited on 01/02/2018).
[22] *OpenOffice.* www.openoffice.org/ (visited on 02/02/2018).
[23] *PAFEC & PafLS.* www.pafec.info/pafec/ (visited on 31/01/2018).
[24] *Room EQ Wizard Room Acoustics Software.* www.roomeqwizard.com/ (visited on 01/02/2018).
[25] *VACS.* www.randteam.de/VACS/Index.html (visited on 02/02/2018).
[26] *WinISD speaker designing software for Windows.* www.linearteam.org/ (visited on 31/01/2018).

PART II

Basic Theory

CHAPTER 1

How Does a Loudspeaker Work?

This chapter will describe the basic operation of a conventional moving coil loudspeaker, beginning with the Rice & Kellogg loudspeaker patent upon which so many of our current speaker designs are based. Before getting into the detail, let's quickly run through the essential operation of a moving-coil loudspeaker. (Feel free to skip this chapter if you are familiar with all this stuff.)

Speakers produce sound waves, which are sensed by our ears and interpreted by our brains. Our ears respond to minute and fast air pressure variations that collectively are known as sound. These rapid pressure changes may also be detected by a microphone, which produces a small alternating voltage called a 'signal'. Such an A.C. electrical signal is very different from conventional mains power, as it varies greatly in both amplitude (height) and pitch (frequency) on a moment-to-moment basis.

Unlike mains power, it is also feeble and can do little if any useful work. Before it can do any meaningful work, we first need to store it until necessary and then at some stage reproduce it at a sufficiently high power level to do useful work.

Sound>Microphone>Storage . . . Storage>Amplification>Loudspeaker

Loudspeakers are normally driven by A.C. electric voltages, but the power or work done actually comes from the current. This is the underlying cause of all sorts of ramifications which will show up later.

An alternating voltage or current at its simplest can be thought of as a wave going up and down smoothly as a period of time passes. It is interesting that the material component (i.e. the air molecules) of any wave stay largely static. However, one molecule encourages movement in the molecules next to it, and these rapidly ripple outward affecting elements sometimes a considerable distance away.

The height of the wavelength is described as the amplitude, while the distance between peaks can be described either as time or frequency (its inverse), in which case we need to take account of the speed of wave transmission in the medium through which it is travelling.

It was proven many years ago that any sound or indeed any waveform can be represented by a series of different individual frequencies and amplitudes, and this is the basis for Fourier decomposition.

In our increasingly digital world, these are chopped up—technically speaking they are *sampled* and converted into digital bits of 0s and 1s—while in the analogue world, we do our best to keep them intact.

Whichever, an electric voltage of varying amplitude and frequency will be fed into the loudspeaker. The nature of voltage is best described as it having the potential to make current flow, but left to itself at low frequencies (below 150 kHz), it tends to hang around and go nowhere. There are very low frequency (VLF) 10 kHz–100 kHz and even extremely low frequency (ELF) 3 Hz–300 Hz radio transmissions.

This signal is amplified (i.e. made larger and more powerful by one or more amplifiers) before being fed to a loudspeaker, whose job it is to turn this electrical signal back into the air pressure waves that our ears detect and our brains recognise as sound or music.

The loudspeaker accepts the voltage that it's fed, and by Ohm's law (voltage = current × resistance), converts this into the current that does the actual work.

For a moving-coil loudspeaker and nearly all electrical machines today,

$$\text{Force} = B \cdot l \cdot I, \tag{1.1}$$

where B is the flux density, l is the length of wire in the gap, and I is the current flowing through the wire.

So fundamentally, a loudspeaker is an electrical machine converting an electrical signal into air pressure variations. Which leads us on to the next question, how does it do this?

1.1 Rice & Kellogg Loudspeaker

First, a brief history lesson: It was back in 1925 when two Western Electric researchers, Chester W. Rice and Edward W. Kellogg, patented the modern moving-coil loudspeaker [1]. Most loudspeakers have remained fundamentally the same ever since, and this is the type of loudspeaker on which this book will concentrate.

One can do a lot worse than read the original patent both for its clarity of thought and its beautiful line drawings. See figure 1.1.

1. Conical diaphragm.
2. Supporting ring secured to the outer edge of diaphragm by:
3. A ring of flexible material, such as silk, rubber or thin leather.
4. Inner edge of diaphragm forming a cylinder we would know as the voice coil former.
5. Inner supporting ring to the voice coil former.
6. Outer supporting ring to the voice coil former.

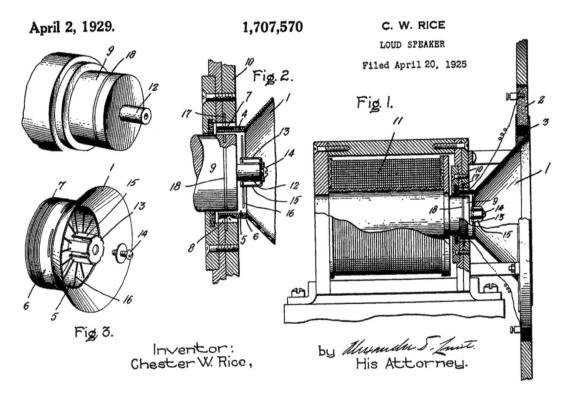

Figure 1.1: Rice & Kellogg Loudspeaker.

7.   Voice coil.
8.   Annular air gap.
9.   Concentric pole piece.
10.   Concentric pole which we would know as the top plate.
11.   D.C. powered coil to produce strong magnetic field.
12.   Extension to pole to locate 13.
13.   A spider supports many arms (15) substantially parallel to the pole.
14.   Fixed by a screw!
15.   These arms being connected by radially extending flexible connecting links.
16.   Flexible links or strands connected to the top of the diaphragm.
17.   Shorting ring to reduce high frequency roll of due to self induction.
18.   Shorting ring to reduce high frequency roll of due to self induction.

Let us see how many of these core components are present in most of our loudspeakers today. The magnetic power is now typically provided by a permanent magnet, numbers 15 and 16 are provided by a suspension or damper often called a 'spider', while numbers 17 and 18 are only rarely used, primarily because of cost concerns. Not bad for a design that's more than 90 years old.

1.2 Left Hand Rule

The principle of operation of such a loudspeaker is as simple as it is elegant. An alternating current flows through a voice coil of many turns, producing a concentrated A.C. magnetic field.

The voice coil is suspended perpendicular (at 90°) to a constant magnetic field. The direction of the resulting force is described by the left hand rule[1] as shown in figure 1.2.

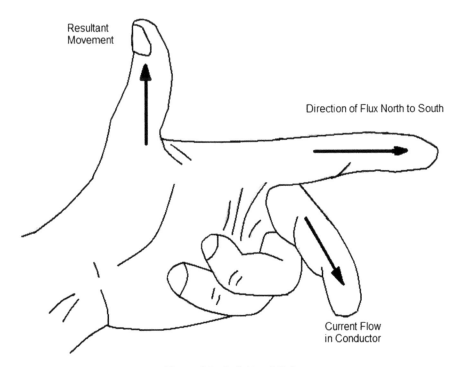

Figure 1.2: Left Hand Rule.

The interaction of these opposing magnetic fields causes the voice coil to move forwards and backwards as the alternating magnetic field generated by the current pushes against the constant magnetic field.

The voice coil is in turn coupled to a diaphragm (originally cone-shaped paper) that vibrates in sympathy with the voice-coil movement. The vibration of this diaphragm in turn vibrates the air molecules, causing the minute air pressure variations that our ears detect and our brains recognise as sound.

Note

1. The left hand rule applies to motors, whilst the right hand rule applies to D.C. generators; but as we are using A.C. signals it makes little difference to the final resulting vibrations.

Reference

[1]   Chester W. Rice and Edward W. Kellogg. *Loud Speaker*. US Patent 1,707,570. Apr. 1929.

CHAPTER 2

Frequency Response

This chapter will outline some of the fundamental methods underlying sine waves, how we calculate the overall level of a signal which is constantly changing over time, and how we display and interpret a frequency response. We'll assume that you have heard the term frequency response, but do not fully know why it is important—and it is important. To do this, we are going on a whirlwind tour of the 'jargon', hopefully in a way that most people can understand.

What we describe here applies equally to sound, light, electrical, or physical waves (like those seen in water). Let's start off with a bath of water. Everyone is familiar with how the water behaves when pushed backwards and forwards in the bath. Keep doing it and the water forms natural waves that quite easily go over the edge of a bath. Believe it or not, this is the key to understanding any general wave behaviour.

If water is pushed backwards and forwards, it is forced into behaviour called a wave. The physical distance between one peak to the next peak of the wave is called the wavelength. The Greek symbol λ is normally used, and the wavelength is measured in metres (or fractions of a metre). We now need to know how fast a wave travels in water (or any other medium). (How quickly something travels is called its velocity.)

The amount of time between one peak and the next peak in the bath could be anything from one to three seconds. For most sound waves, however, the time is measured in intervals of a few ms or μs (milliseconds or microseconds).

From the wavelength and velocity it's possible to work out the frequency, as frequency equals velocity divided by wavelength. However, we also need to know something about the height of the waveform, also called the amplitude.

With a bit of mathematical trickery it's possible to work out the effective equivalent level. Indeed, any regular repetitive waveform can define a level that represents its amplitude.

Furthermore, when pushing the water in our virtual bath backwards and forwards, it would sometimes go higher and higher. This is called resonance, and we will come to it later as it describes some other important bits of loudspeaker behaviour.

2.1 Generating a Sine Wave

We were admonished by one reviewer for skipping over what a sine wave was and how it is represented. A sine wave can be generated from a circle, and electrical generators produce sine waves by just this method. This is shown graphically in figure 2.1.

Let's start with a circle at zero degrees. At each interval of a few degrees, draw a horizontal line and plot its intersection with the interval seen on the time axis. So we set zero degrees at zero time, 15 degrees at time interval 15, and so forth.

As we move in the direction of the arrow, we will move at a constant rate in this case, so we will produce a single fixed frequency. Figure 2.1 also shows clearly the intertwined relationship of frequency and time.

The second half of figure 2.1 shows an amplitude versus time graph. Notice how a sine waveform is drawn as we rotate around the circle, so just speed up the rate of rotation until we reach the desired frequency range.

The next point is that lots of people describe a loudspeaker as having such-and-such a frequency response. It is fundamental to our understanding that the description is clearly understood.

We said earlier that some mathematical trickery was needed to work out the effective total level. Looking again at a sine wave, a positive half is followed by a negative half, and one technique is simply to flip one half over to the other half as shown in figure 2.2. We can then calculate the area under both halves of the curve and add them without cancellation.

The term RMS stands for root mean square, which is the mathematical process that is often used. What we actually do is square each individual value (multiply it by itself); this changes any negative numbers to positive ones. We then sum (add together) all of the values over the time period we are interested in. Finally, we take the square root of this summed value—giving us an RMS value. One might also find 'peak-to-peak' (especially in digital systems), 'peak', or 'average' values, but please be careful about the units that are being used.

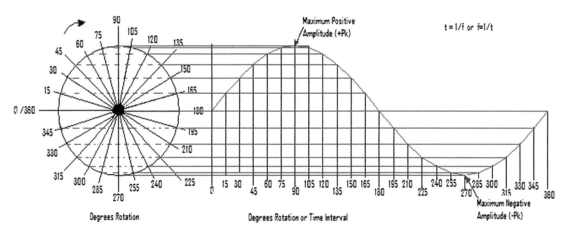

Figure 2.1: Generating a Sine Wave.

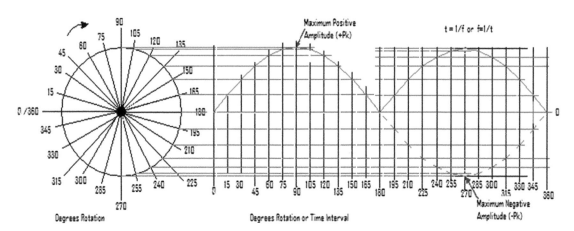

Figure 2.2: Calculating a Sine Wave's Amplitude.

Simply put, we then draw, plot, or predict a graph of this total level, however it has been calculated, and do so against each frequency point in turn. At this point be grateful that our computers and software are doing all this, as it would be really tedious to do it manually.

2.2 Practical Swept Sine Waves

Let's look at a real example now: Taking a swept sine as produced by a Klippel analyser as shown in figure 2.3, we will look at how it changes over time. At low frequencies, it looks quite like our sine wave from earlier.

Now let's take the same swept sine wave as produced by a Klippel analyser as shown in figure 2.4. We will look at amplitude vs frequency (Hz) as a series of spectral lines.

Looking at this we can easily see two important things:

- Amplitude is higher at low than at high frequencies.
- It appears to have many gaps at low frequencies while it appears to be a solid mass at high frequencies.

So what's going on? Why has a nominally flat series of sine waves changed so drastically? Quite simply, there are many more frequencies in any given period of time at higher frequencies than at lower frequencies.

So the total amplitude is flat, and we can see it on the graph—the spectral balance slopes downward at −6 dB per octave.[1] Please remember that this latter curve is a mathematical representation, as it does not actually exist, whereas the first curve is an actual representation of the instantaneous voltage at a particular point in time.

However, on a more practical level we need to measure things like frequency response. Let's apply our first signal, the swept sine wave, to a real loudspeaker and have a look at the results step by step. We'll look at the voltage output from a microphone as measured in an anechoic chamber.

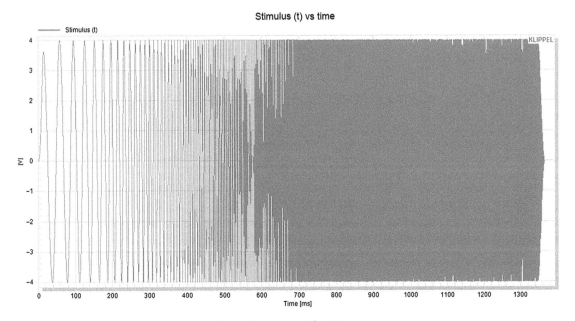

Figure 2.3: Swept Sine Wave.

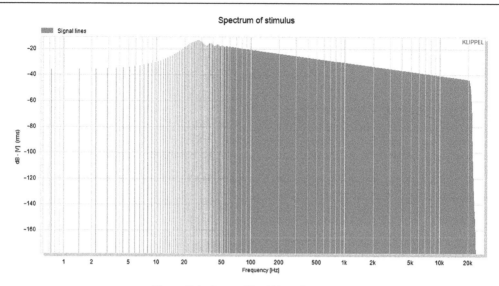

Figure 2.4: Swept Sine Wave Spectrum.

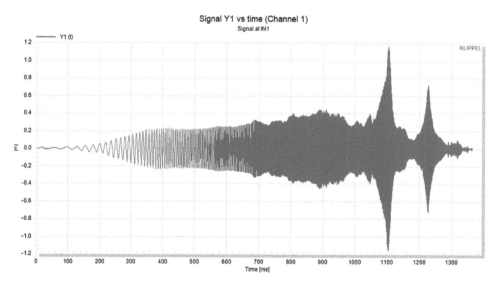

Figure 2.5: Swept Sine Wave of a Loudspeaker.

Let us take the same swept sine wave, but this time we will look at it as the signal produced by a loudspeaker. This is shown as figure 2.5, and we can see immediately that there is tremendous variation at different frequencies, and also that it is mirrored about the zero volt axis and looks very peaky at high frequencies.

Next, we will take the same swept sine wave as shown in figure 2.6. However, this time we will look at it as a series of spectral lines—amplitude versus frequency (Hz). Technically we will run a fast Fourier transform (FFT) at a series of frequencies, the sampling frequencies, and display the amplitude of each of these individual frequency bins in turn.

We can begin to make out a more or less conventional shape now. However, as this would still be difficult to interpret, we divide the output signal by the input signal, as we usually make comparative measurements.

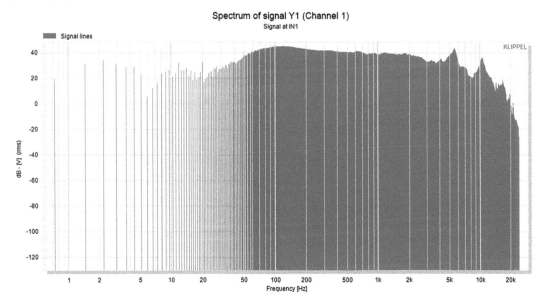

Figure 2.6: Swept Sine Wave—Loudspeaker Spectrum.

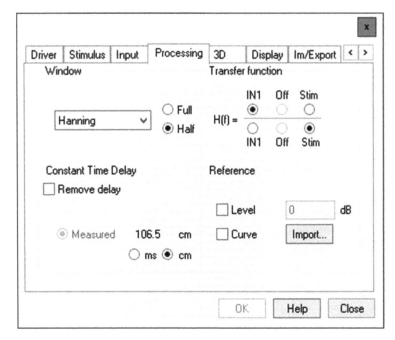

Figure 2.7: Transfer Function.

This then removes the apparent amplitude slope shown; this is called a transfer measurement or magnitude of response. This magnitude or loudness (hence sound pressure level or SPL) is normally expressed in decibels (dB). We normally use the ratio of the microphone output signal divided by the input signal. This is shown in figure 2.7.

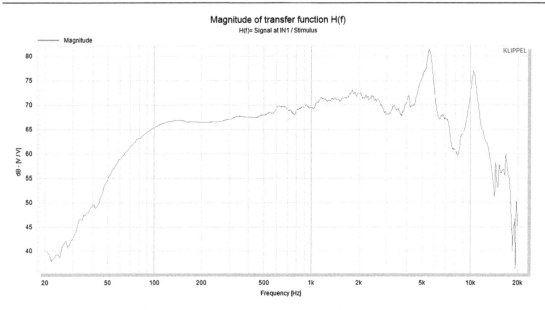

Figure 2.8: Swept Sine Wave—Loudspeaker Spectrum.

Now, as the input and output are both synchronised with each other, we are left with a series of points, and we simply draw or plot a curve in between successive points to give what is called a frequency response curve. Note, however, that as we normally use a logarithmic amplitude scale or dB, we display this using a linear vertical scale or Y-axis, whilst conventionally we use a logarithmic scale for the frequency or X-axis, as this fits closer to the ear's resolution.

The same swept sine wave transfer response, as produced by a Klippel analyser, is shown in figure 2.8, and we will look at this series of joined amplitude points versus frequency (Hz) as an amplitude versus frequency response—often just called a response curve.

As we said earlier, convention dictates that we usually plot the amplitude or sound pressure level (SPL) on a linear scale. However, the overall level calculations are made on a logarithmic scale (e.g. decibels SPL), while the frequencies are plotted on a logarithmic scale.

This means that considerable care is needed when adding together SPLs, ensuring that you are using the correct units. As with logarithmic units like SPL, simply adding the values is equivalent to multiplying the levels together, so please be careful. (For further information on handling logarithmic units please look it up online or see a mathematical text book like *Engineering mathematics* [1].)

Note

1. An octave is a doubling or halving of a frequency.

Reference

[1]   Kenneth Arthur Stroud and Dexter J. Booth. *Engineering mathematics*. Palgrave Macmillan, 2013.

CHAPTER 3

Frequency Versus Time

This chapter will start by outlining some of the differences between a frequency response and the time domain.

As demonstrated earlier when we produced the sine wave, the frequency domain is inherently tied to time. There is a direct correspondence between the two involving angular rotation, π and the S-plane, which we will not go into here.

Formally speaking, we translate between the time domain and the frequency domain by using the Fourier transform; the inverse Fourier transform goes between the frequency domain and the time domain.

Many such analyses are thus said to be in the 'time domain', and here the system is described in terms of amplitude, taking account of sign +/− and time.

The important point is that we can describe a system or a loudspeaker equally using either domain and can freely change between the two with the appropriate tools. However, some aspects of design and performance are clearer and more easily calculated in one domain, whereas others are easier to handle in the other domain.

From the theoretical perspective, Richard Heyser's many papers contributed greatly from 1967 onward, and many of these were published by the Audio Engineering Society (AES) who produced a combined anthology of many of his works in 1988 [4].

The practical use of many of these techniques probably first came to the notice of the audio loudspeaker community with the publication of Laurie Fincham and Michael Berman's paper 'The Application of Digital Techniques to the Measurement of Loudspeakers' in 1977 [1].

What aspects? Obviously those which change over time are clearer in the time domain. These include such things as damping or resonance, and the step response, together with so-called 3D responses or waterfall curves (being a blend of both domains as they are composed of a series of frequency responses but spaced out in discrete time intervals).

Depending upon the methods used to generate them, further analysis may be possible. For example, those created from a maximum length sequence (MLS) can measure the noise floor. A major leap forward was provided by the production of the DRA MLSSA System [6] and the publication of Douglas Rife and John Vanderkooy's paper 'Transfer-Function Measurement with Maximum-Length Sequences' [7]. By changing to a swept sine wave and utilising the log sweep techniques of Farina [3] we can also measure harmonic distortion and total harmonic distortion (THD).

The impulse response also allows us to select certain portions of it for analysis while leaving others. This process is called *windowing*. In principle it is simple, but like any technique there are many other things to consider, from the shape of the window to the signal under consideration.

3.1 Perfect Impulse

Strictly speaking, of course, a perfect impulse response does not and cannot exist. It is better described as an impulse waveform—and by definition a waveform is a realisable form rather than an arbitrary abstraction. (A closer realistic approximation is the Kronecker delta which is mathematically related to the Dirac delta function.)

For an example see figure 3.1, which is from the calibration curve of HOLMImpulse [5] taken at 96 kHz. A single positive pulse that returns to zero without either over or undershooting can clearly be seen. (The frequency response is a perfectly straight line from 1 Hz to 100 kHz—in theory it should be from 0 Hz to infinity!)

So what can an impulse do for us that a frequency response cannot?

1.   We can determine how a system performs with respect to time.
2.   We can 'isolate' or examine various time sections; for example, harmonic distortion gets separated out to much later time periods.
3.   We can determine 'damping'; ringing and resonances especially can show up very clearly in the time domain.
4.   We can ignore certain artefacts earlier or later in the time record, as these correspond to reflections.

From the 'perfect impulse' response we can see very little, so let us deliberately introduce some problems that we will see later.

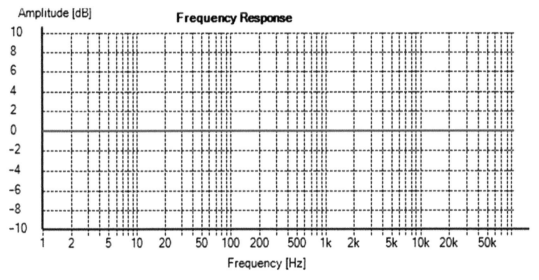

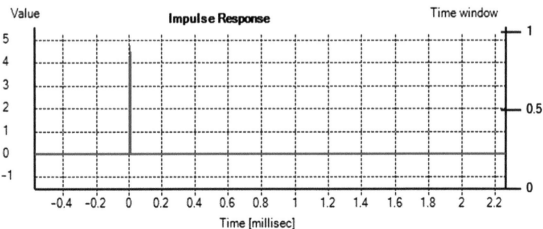

Figure 3.1: Perfect Impulse.

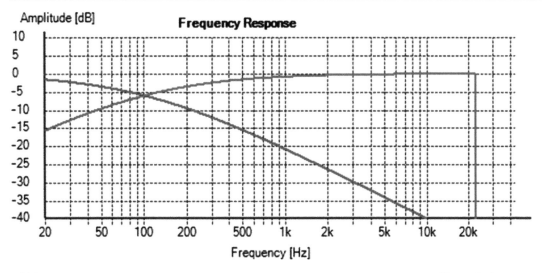

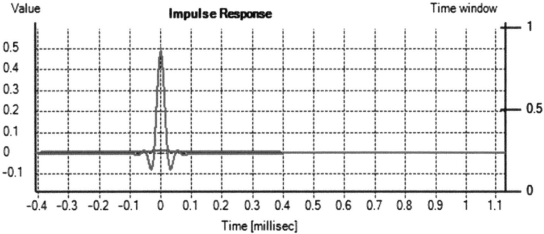

Figure 3.2: −6 dB at 100 Hz.

3.2 100 Hz 6 dB/Octave Filters and Impulse Responses

First, all conventional loudspeakers have a roll-off at low frequencies, so let us introduce a simple filter: −6 dB at 100Hz with a slope(s) of −6 dB/octave above and below 100 Hz. This is shown as figure 3.2.

Immediately we can see a dramatic change in the frequency and phase responses as well as in the impulse versus time.

3.3 Midrange Driver

Next we look at a midrange driver at a relatively high SPL as measured by the CLIO Pocket measurement system [2]. First, examine the frequency response versus sound pressure level (decibels SPL); this is shown as figure 3.3.

We will now look at a waterfall curve of the same loudspeaker driver; this can be seen as figure 3.4.

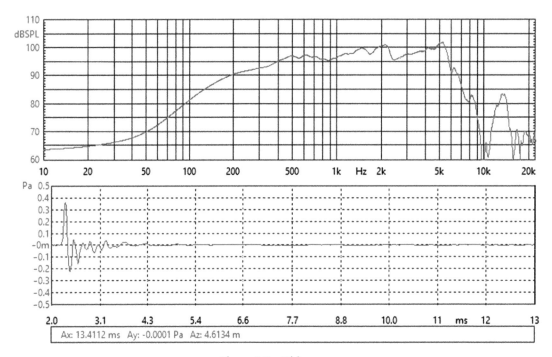

Figure 3.3: Midrange.

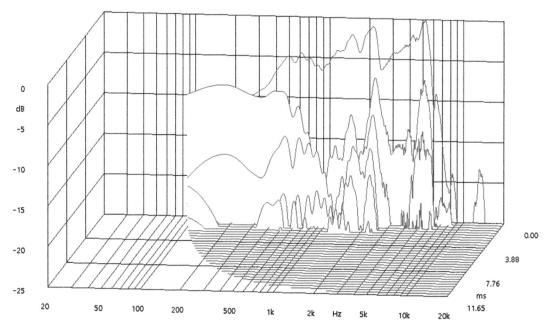

Figure 3.4: Midrange Waterfall Plot.

Here we can clearly see the change in the frequency response versus time slices. A waterfall plot typically does this at multiple time intervals and can reveal much information about a loudspeaker or system and especially about any resonant behaviour.

References

[1]   J. Michael Berman and Laurie R. Fincham. "The Application of Digital Techniques to the Measurement of Loudspeakers". In: *J. Audio Eng. Soc* 25.6 (1977), pp. 370–384. www.aes.org/e-lib/browse.cfm?elib=3366.

[2]   *CLIO Pocket*. www.audiomatica.com/wp/?page_id=1739 (visited on 01/02/2018).

[3]   Angelo Farina. "Simultaneous Measurement of Impulse Response and Distortion with a Swept-Sine Technique". In: *Audio Engineering Society Convention 108*. Feb. 2000. www.aes.org/e-lib/browse.cfm?elib=10211.

[4]   Richard C. Heyser. *An anthology of the works of Richard C. Heyser on measurement, analysis, and perception*. Audio Engineering Society, 1988. www.aes.org/publications/anthologies/downloads/jaes%5C_time-delay-spectrometry-anthology.pdf (visited on 04/02/2018).

[5]   *HOLM Impulse*. www.holmacoustics.com/holmimpulse.php (visited on 31/01/2018).

[6]   *MLSSA Acoustical Measurement System*. www.mlssa.com/ (visited on 04/02/2018).

[7]   Douglas D. Rife and John Vanderkooy. "Transfer-Function Measurement with Maximum-Length Sequences". In: *J. Audio Eng. Soc* 37.6 (1989), pp. 419–444. www.aes.org/e-lib/browse.cfm?elib=6086.

CHAPTER 4

Resonance and Damping

4.1 Resonance

In this chapter we describe resonance: what it is, some of the effects it has on a system, and how we can calculate it from knowing the mass and compliance.

What is resonance? Simply put, resonance is the tendency of a system or part to oscillate or to keep vibrating more at some frequencies than others. Why does resonance matter? Depending upon the enclosure type, it has a critically important role to play as it often sets the low-frequency limits to which the combined driver and enclosure will go.

Most loudspeakers are inefficient machines, and a key decision is to operate most loudspeakers in the mass controlled range, using a conventional cone. This was probably the core difference between the Rice & Kellogg loudspeaker [4] and Jensen's earlier design [3].

With this type of design, the displacement falls off with increasing frequency at –6 dB per octave, at the same time as the effective load is increasing at +6 dB per octave. The result is a flat frequency response over a given frequency range. Thus, one of the main tasks in conventional loudspeaker design has been laid bare—gaining a sufficiently wide bandwidth for a given loudspeaker's sensitivity or efficiency, given the prevailing cost and size constraints of the individual project.

Very often, one of the most important decisions in drive unit design is actually the resonant frequency.

Essentially, mechanical resonance is related to two things: compliance (how easily something moves) and mass (how heavy it is). The system resonant frequency F_s (as it is usually known) is calculated from equation 4.1:

$$F_s = \frac{1}{2 \cdot \pi \cdot \sqrt{C_{ms} \cdot M_{ms}}} \tag{4.1}$$

where C_{ms} replaces the capacitance and M_{ms} replaces the inductance in the equation.

4.2 How Can We Measure Resonance?

At its simplest we just need to see it. We do this by plotting whatever parameter is varying, usually against frequency: the peak or peaks are normally resonant frequencies or modes. Also, very often the precise resonance frequency is best shown by a zero phase.

Normally, for a loudspeaker, this is measured in free air. It is important to do this, as any significant reflections from walls or other objects can make a significant difference to the results. However, very few modern loudspeakers are actually designed to be operated like this and may easily be damaged by operating at high levels in free air or unloaded conditions.

One of the most inaccurate ways is to sit a loudspeaker driver on a flat hard bench, on its magnet and with the diaphragm pointing up into the air. See figure 4.1.

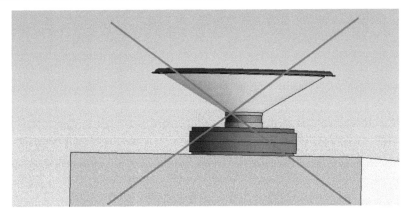

Figure 4.1: Loudspeaker on a Bench.

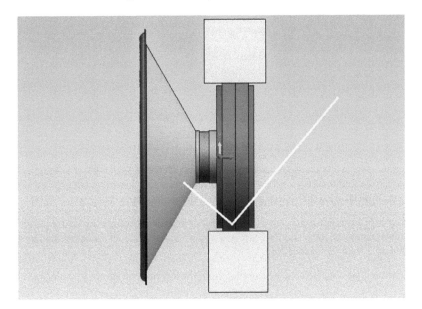

Figure 4.2: Resonance of a Loudspeaker on a Bench.

This will cause problems as follows: There will be a strong reflection from the bench onto the rear of the diaphragm; there will be an extra air mass due to the air between the rear of the diaphragm and the bench; and gravity will pull the diaphragm down towards the bench. Different problems will occur when measuring the driver in an enclosure.

So what should we do? Generally speaking, it is desirable to ensure the following: that the loudspeaker is operating in its normal orientation; that it is operating in free air; that the loudspeaker is not near to surfaces, as these can reflect sound pressure back; and that the measurement is made in a reasonably quiet environment.

Klippel recommends the use of a dedicated speaker frame [1] rigidly holding a conventional loudspeaker driver in a vertical orientation. An example is shown as figure 4.2.

It is essential to avoid sudden changes of air pressure, easily caused by closing doors. It is usual to take the measurements at small signal levels, and for horn loudspeakers or micro speakers it is often done in a vacuum.[1]

4.3 Damping

What is damping? Resonance is the tendency of a system (or loudspeaker) to oscillate or keep vibrating more at some frequencies than others. Damping acts to control or stop this tendency and usually acts with respect to time.

This chapter will examine both electrical damping (as applied to the measurement of electrical Q) and also acoustical damping and the effect of a lack of it.

Damping obviously has the effect of slowing down an oscillating system. However, it can also prevent a system from responding fast enough to a desired change. Damping usually shows its effects most clearly in the time domain.

Do we need damping? Its absence in a metal dome tweeter can make a sound like a pair of dustbin lids being banged together! And a cymbal with too much damping would not sound like a cymbal.

So how can we measure damping? Damping can be measured in both the frequency and time domains.

In the frequency domain, we measure amplitude and phase versus frequency, whilst in the time domain we measure the change or decay of the envelope (or the instantaneous value) versus time.

From either domain it's then possible to extract the required information and apply the necessary calculations.

We will concentrate on the frequency domain and the case of electrical damping, where one often measures the electrical impedance in order to calculate many important loudspeaker parameters that will come in later.

4.4 Measuring Electrical Damping

As we saw earlier, from an electrical signal, Ohm's law can be used to find a resistance: as $V = I \cdot R$ so $R = V/I$.

This may be done either using a current source (essentially putting a large enough resistor in series with the loudspeaker) or as a voltage source (using a very small series resistance and then measuring the current directly or calculating the current by measuring the voltage across a small sensing resistor).

Either way, we end up with a transfer function voltage/current or impedance (ohms) versus frequency (Hz). If possible we would always measure the phase (in degrees) versus frequency (Hz) at the same time, as very often the phase shows clearer details than the impedance.[2]

Once we have these curves, we can calculate the damping from the information shown in figure 4.3.

First measure the maximum value Z_{max} and its frequency F_0. Then measure the minimum value[3] Z_{min}. Then we can calculate

$$Z_{mean} = \frac{Z_{max}}{Z_{min}}. \tag{4.2}$$

In this case, $Z_{max} = 48.478$ and $Z_{min} = 3.647$, so $Z_{mean} = 13.292$.

Now we need to find the frequencies F_1 and F_2 where the value equals Z_{mean}, as shown in figure 4.4.

Now calculate

$$Q = F_0 \cdot \frac{Z_{mean}}{F_2 - F_1}. \tag{4.3}$$

In this case, Q approximately equals $41.347 \, \text{Hz} \times 13.292/(127 - 12) \approx 4.75$.

Basically the same procedure applies for mechanical vibration measured by a laser: take readings of the amplitude in all cases and for Z_{min} use the value at a low frequency away from the resonance.

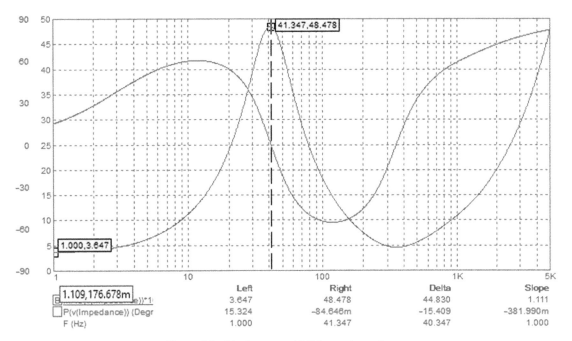

	Left	Right	Delta	Slope
B MAG(v(Impedance))*1	3.647	48.478	44.830	1.111
P(v(Impedance)) (Degr	15.324	-84.646m	-15.409	-381.990m
F (Hz)	1.000	41.347	40.347	1.000

Figure 4.3: Maximum and Minimum Impedance.

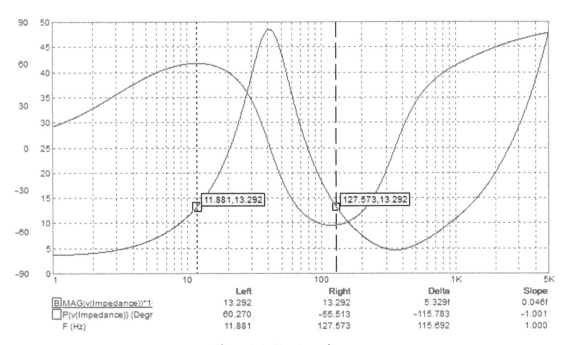

	Left	Right	Delta	Slope
B MAG(v(Impedance))*1	13.292	13.292	5.329f	0.046f
P(v(Impedance)) (Degr	60.270	-55.513	-115.783	-1.001
F (Hz)	11.881	127.573	115.692	1.000

Figure 4.4: Z_{mean} Impedance.

As a check to ensure you have a true resonance, in most cases the phase should transit through a phase zero as we saw earlier. Also, the result of the calculation

$$F_s \approx \sqrt{F_1 \cdot F_2} \tag{4.4}$$

should also be approximately equal to the resonance frequency.

4.5 Acoustical Damping

Many years ago at Goodmans, we had a problem related to damping, though we did not know it at the time. We were working on a hi-fi loudspeaker using a Cobex cone.[4] It measured really well but sounded terrible, as we had neglected to look at the time domain!

Back then in around 1990, MLSSA [2], one of the first PC-based FFT analysis systems, had just come out and Goodmans had just bought one; a quick look at the time domain of the loudspeaker (sorry, but the data is long gone) revealed that the speaker was severely smeared in the time domain, with very poor decay.

We certainly had no effective modelling capability back then, but Goodmans had just invested heavily in PAFEC, and we were beginning to start modelling. Although we had been on the course with Patrick Macey at PAFEC, the software was then in the hands of a student who was working off site. A solution was needed immediately. However, we did have access to a laser vibrometer; could this give answers to the problem?

We reasoned that if the whole cone and loudspeaker had this delayed response then it must be the cone material that was causing the problem. We therefore did not need to analyse the whole cone. How would a strip of material respond when it was shaken? It was simple enough to build up a special motor unit and point the laser beam at a white spot on the end of a beam of material 5 mm × 50 mm or so.[5] A rough sketch is shown as figure 4.5.

We then connected up MLSSA to provide the signal and analysis. A few minutes later—remember this was in the days of using a 386 laptop fitted with a 387 co-processor—and we had the initial results.

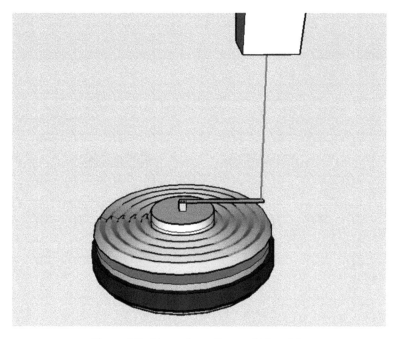

Figure 4.5: Measuring a Beam of Material.

The results imported by VACS [5] as shown in figure 4.6 showed the problem—it was ringing for over half a second! An eternity for a loudspeaker.

The solution was simple. BBC engineers had found out earlier in the decade that applying some dope (damping material) to the back of the beam (and subsequently the cone) would be effective. The result in the time domain is shown as figure 4.7. Clearly, the ringing has been effectively stopped with no trace of the prior problem remaining.

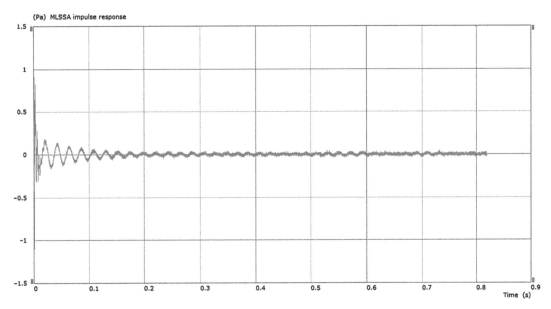

Figure 4.6: Impulse Response of Cobex Material.

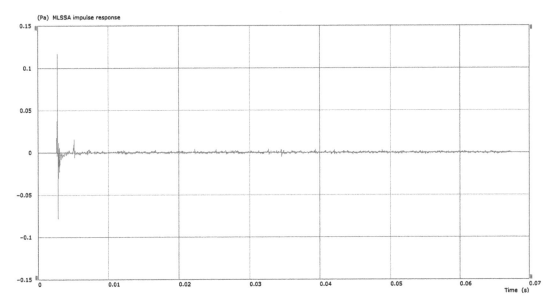

Figure 4.7: Impulse Response Final Loudspeaker.

The result was a tremendous improvement in clarity, especially on vocals, though it hardly affected the frequency response, and the extra mass lost perhaps 0.25 dB of sensitivity.

This shows how acoustic damping or its lack can affect an otherwise fine loudspeaker. It also demonstrates how we can prevent such problems from getting started with our modern tools and simulations.

Notes

1. The key insight that vacuum measurements give is the separation of mechanical movement from acoustical loading conditions and the additional information this gives a designer.
2. This can be clearly seen in the graph as the phase usually goes through zero degrees.
3. Often taken as the D.C. value; this is *not* accurate as it does not include the effects of damping, which often cause the value to rise.
4. We can hear the groans of more experienced engineers.
5. The white spot was created using a typing correction pen.

References

[1]   *Klippel Pro Driver Stand.* www.klippel.de/fileadmin/klippel/Bilder/Our_Products/ R-D_System/PDF/A5_Pro_Driver_Stand.pdf (visited on 04/02/2018).
[2]   *MLSSA Acoustical Measurement System.* www.mlssa.com/ (visited on 04/02/2018).
[3]   Edwin S. Pridham and Peter L. Jensen. *Telephone.* US Patent 1,105,924. Aug. 1914.
[4]   Chester W. Rice and Edward W. Kellogg. *Loud Speaker.* US Patent 1,707,570. Apr. 1929.
[5]   *VACS.* www.randteam.de/VACS/Index.html (visited on 02/02/2018).

CHAPTER 5

Finite Element Analysis

5.1 What is FEA?

This chapter describes in very simple terms, and without the mathematics, what finite element analysis (FEA) is about and how it achieves its results. We discuss what is important in any model and how in our view it is essential to be modelling the most important things or parts first. In this book we are not concerned with all of the various types of analysis and the many mathematical methods used in solving these; suffice to say they all have advantages and disadvantages dependant upon the problem, the resources, and the complexity of the calculations required. Rather, throughout this book we are concerned with the process of design and modelling in practice.

Many problems in physics and engineering are governed by various equations that describe the changes of numerous variables. For example, in loudspeakers, flux, displacement, and pressure are key variables that must be simulated.

One approach to solving these equations is the finite element method (FEM). Taking the example of a substantial acoustic domain, the pressure distribution is very complex throughout (so a complicated Helmholtz equation would be required). However, within a very, very small neighbourhood, the pressure variation is much, much simpler.

So instead of analysing the whole region, we will instead break or decompose the region into smaller and simpler shapes. To keep things really simple we will assume a 2D flat area which we will break up with triangles that are small enough so that within each triangle the pressure distribution can be approximated by a linear variation. In this case, the FEM solves the Helmholtz equation exactly, assuming that the variation is linear within each triangle.

FEA thus effectively swaps the mathematical complexity of a single region for multiple but much simpler equations, each covering much smaller regions—albeit ones that are solved simultaneously and must be performed many times; each time these calculations are made they are known as the 'degrees of freedom'.

The price that must be paid for this simplicity is that an extremely large number of calculations need to be made and the results stitched together. Fortunately, computers are very good at doing large numbers of simple calculations very quickly, so a methodology now exists that is capable of solving complex problems.

From the user's point of view, we do not need to know how these programs do their jobs any more than a driver needs to be able to design a car (or the parts and components therein).

We can use the results, but it is necessary to have a feel for them and any possible problems, so that we do not get misled by our technology, in just the same way that a driver using a satellite navigation aid would be well advised to check before going down a small path leading to a mountain. There may be a path through to where he wants to go, but not necessarily one that a vehicle can get through. It's a poor analogy perhaps, but it may be useful. If the results seem too good to be true then perhaps they are.

Generally speaking, there are two types of finite element programs: specialist and general purpose ones. Currently the best of both are easier to use than earlier examples, and the specialist ones are generally easier or simpler to set up and use, whereas the more general ones tend to be more complicated. One thing separating the two is the amount of knowledge and data that one has to enter.

There is a saying: 'All models are wrong but some are useful.' We firmly believe this should be a guide with any type of model, but even more so with FEA. Remember FEA is a modelling technique, and as such it is not and cannot be *totally accurate*. However, we can make it accurate enough for the job in hand.

An essential thing here is not trying to model everything; rather, keep to an appropriate level of detail. Look at it this way: If we had to make a model of the world, what would be the most important thing to model? Let's assume we were interested in modelling the weight (or more correctly the mass) of the world, taking everything into account. What, very roughly, would each component part of such a model weigh?

- A microbe $= 10^{-12}$ kg or approx 0.000,000,000,001 kg.
- An individual person $= 10^2$ kg or approx 100 kg.
- A mountain, say Mount Everest $= 10^{12}$ kg or approx 10,000,000,000,000 kg.
- Or the whole globe $= 6 \times 10^{24}$ kg or approx 6,000,000,000,000,000,000,000,000 kg.

Obviously, in this case, the whole globe would dominate a model of the world, as even the tallest mountain—let alone a person or a microbe—is insignificant in comparison.

What has this got to do with modelling a loudspeaker? It's very tempting to take a full CAD model and build an FEA model on this. In fact, many FEA and boundary element method (BEM) programs include import capabilities from various CAD and 3D CAD formats. Although it's tempting to use these, we would strongly advise against it.

Especially if you are new to loudspeakers, there might be an understandable tendency to include everything by having as detailed a model as possible, the rationale being to try to increase accuracy.

Surely the aim of modelling is complete accuracy? We would argue differently: particularly in the beginning of modelling, the converse is true, as many things are still unknown. This changes when you have an established model working, as often then the focus does change to absolute accuracy.

Our reasoning is that the full 2D or 3D CAD model and drawings will inevitably have much detail, dimensions, perhaps tables, and other relatively unimportant items in them. This complexity can mean that lines do not meet accurately, or a lot of clean up is necessary, or modifications are required to make the model work efficiently. So sometimes the FEA model will need to be optimised by removing sharp corners and edges to allow the meshing process to function correctly. So simplify, simplify, and simplify is one of the key concepts in FEA.

We would therefore strongly recommend either constructing a new FEA drawing on a node-by-node basis, or that you use a simple and clean *.dxf file [1] to build the model. This might perhaps even be generated from an FEA package like FEMM [2]. Later on, after the FEA analysis is completed, it can be exported into the main CAD drawing(s).

5.2 Uses for FEA

In this book, we will be using a magnetic FEA program (FEMM) [2], mechanical FEA programs Mecway [5] or Fusion 360 [3], and also a mechanical–acoustical program (PafLS) [6]. The key point here is that FEA is a computational technique to aid our calculations, but to make these calculations the techniques must be applied effectively.

This requires knowledge of the underlying parameters, which may be physical, material, acoustical, electrical, optical, magnetic, or whatever domain we wish to model. So depending upon the program or programs, one of our first tasks is to gather such information.

5.2.1 Magnetic FEA

As an example, FEMM has a reasonable internal database of many of the things it needs to know: wire gauges, materials, conductance, magnetic materials, and so forth. So it's relatively painless to get useful results quickly.

5.2.2 Electromechanical to Acoustical FEA

Likewise, PafLS has an internal database of materials used for its mechanical calculations, while taking as inputs various other parameters that are fed into predefined templates.

5.2.3 Mechanical

More general purpose programs like Mecway or Fusion 360 do not have access to equivalent data. So depending upon the program, you may well need to find out the missing information. This is not always easily available so a certain amount of research may be required. Fortunately the internet is very helpful here, especially sites like Wikipedia and materials databases like MatWeb [4].

Often information is freely available, but under a military standard (MIL) or NASA specification name or number, and is often charged for privately by various organisations. However, even the best efforts may not be able to find the information, so it may then need to be measured somehow.

The point here is that with a bit of effort, application, and research it is quite possible to measure accurately even quite esoteric physical quantities as, for example, Young's modulus, which is covered in the next chapter.

References

[1] *AutoCAD 2012 DXF Reference.* http://images.autodesk.com/adsk/files/autocad_2012_pdf_dxf-reference_enu.pdf (visited on 04/02/2018).
[2] *Finite Element Method Magnetics.* www.femm.info/wiki/HomePage (visited on 31/01/2018).
[3] *Fusion 360 ™ 3D CAD, CAM, and CAE.* www.autodesk.com/products/fusion-360/free-trial (visited on 31/01/2018).
[4] *MatWeb, Materials Information.* www.matweb.com/ (visited on 31/01/2018).
[5] *Mecway Finite Element Analysis for Windows.* http://mecway.com/ (visited on 31/01/2018).
[6] *PAFEC & PafLS.* www.pafec.info/pafec/ (visited on 31/01/2018).

CHAPTER 6

Young's Modulus

The mechanical properties typically used in FEA are used to model the mechanical and ultimately the acoustic behaviour of loudspeakers.

However, before doing so we need to divide our materials into two major classes: *isotropic* and *anisotropic*.

Isotropic materials have constant material properties in all directions. Therefore they have nominally consistent properties in any and all directions. Examples include most metals, expanded foams, and glass.

Anisotropic materials have properties that vary significantly with respect to direction. Examples are some extrusions, carbon fibre filaments, honeycomb assemblies, much wood, bones, and many natural cellular structures.

6.1 Young's Modulus

Young's modulus is the ratio of stress (force per area) to strain (the ratio of new size to original size). We will see how it can be measured and find why it is so important to modern modelling techniques.

Many different variants of Young's modulus exist, depending upon how a material is manipulated. The Young's modulus of a material is only valid when it is still in an elastic condition so that it obeys Hooke's law [2].[1]

Hooke's law states that: *The force F needed to extend or compress a spring by some distance X is proportional to that distance.*

$$F = kX \qquad (6.1)$$

where k is stiffness (a constant characteristic of a spring), and X is small compared to the total possible deformation of the spring.

Young's modulus measures the amount of force versus area required to extend or compress a material's shape. The point we need to bear in mind is that when any material is extended or compressed, it also becomes more or less flexible in the process. So as the material becomes thinner or thicker dynamically, this results in a change in the shape of, say, a loudspeaker's diaphragm—and in turn the shape supports different modes of vibration.

This is of particular importance as it means that a loudspeaker's diaphragm could have a different shape at different frequencies. And a differently shaped diaphragm will result in a different frequency response.

Young's modulus is important because it allows modern calculations (especially finite element methods) to directly determine how a material responds under both static and dynamic conditions. The classical method for measuring it was to measure the elastic stretch of a vertical wire under strain (in tension).

6.2 Measuring Tensile Young's Modulus

This is a very simple experiment and is shown in figure 6.1.

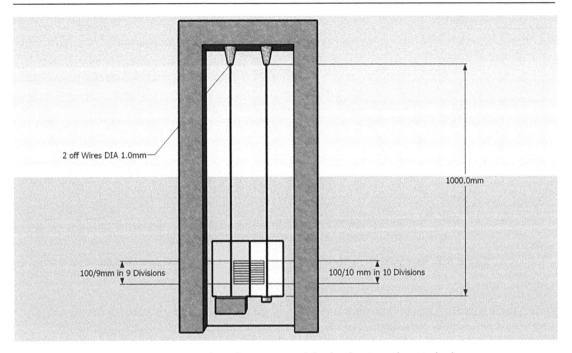

Figure 6.1: Measuring of Young's Modulus by the Extension Method.

To do this we construct a frame and hang two identical wires of say 1.0 mm diameter from it. We keep one wire lightly tensioned with a small weight such that the force is *not* enough to extend its length significantly, whilst we hang a heavier weight from the other wire—note it is *essential* that this weight must not be so heavy that either wire is stretched beyond its elastic limit, especially at first as then you would be measuring the *nonlinear* range, where recovery is not possible.

By knowing the weights used along with the force of gravity, roughly 9.71 N/m, we then measure the extension of the wire under stress and calculate the tensile Young's modulus as shown in equation 6.2. In the example shown, the two wires have been overlaid by two pieces of paper upon which have been marked a series of grid lines. These form a Vernier scale; this enables us to resolve fine differences in measurements visually.

The Young's modulus (*E*) can be calculated from:

$$E \equiv \text{tensile stress/extensional strain,}$$

so

$$E = \frac{\sigma}{\varepsilon} = \frac{F/A_0}{\Delta L/L_0} = \frac{FL_0}{A_0 \Delta L} \tag{6.2}$$

where:

- *E* is the Young's modulus (modulus of elasticity).
- *F* is the force exerted on an object under tension.
- A_0 is the original cross section area through which the force is applied.
- ΔL is the amount by which the length of the object changes.
- L_0 is the original length of the object.

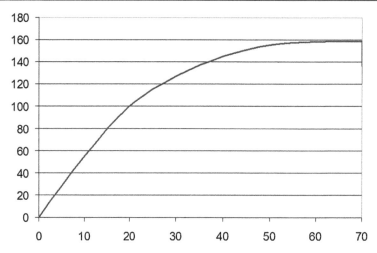

Figure 6.2: Raw Data from Measuring Young's Modulus by Extension Method.

In principle this is relatively simple. However, it is worth taking not one but several measurements, starting with a small weight and gradually increasing it. (You will eventually pass the elastic limit and ultimately exceed its breaking point.)

Plotting all of these figures will show a full Young's modulus curve of the material. This is not usually completely linear except at very low levels, hence the warning to keep to low weights. A typical curve is shown as figure 6.2.

6.3 ASTM E 756-05

A general test configuration was described by Oberst. This is often called the Oberst beam or bar and is a general method of measuring damping using a vibrating cantilever beam. Referred to as the Oberst beam test, it appears in many standards including ASTM E 756-05.

A useful paper is 'Measurement of Dynamic Properties of Materials' [1], see figure 6.3. This was actually the technique we used back at Goodmans to analyse the misbehaving cone material; see figure 4.5 on page 22.

In this method, a beam is rapidly vibrated physically over a range of frequencies. As we are using Young's modulus for loudspeakers we are mainly interested in the dynamic properties of materials, so this method will give us other useful information; however, we still need to be careful as many materials used for loudspeakers are very thin, light, and easy to push beyond their elastic limit.

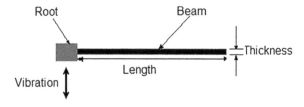

Figure 6.3: Measuring of Young's Modulus by the Oberst Beam Method.

6.3.1 Klippel method

Another variant of this method is to excite a beam of material by using an acoustic rather than a mechanical source. An example of this is that of Klippel's material parameter measurement (MPM) [4]. This measures the vibration of a beam of material using a triangulation laser, with said beam being acoustically excited. The processing is conducted by the MPM module of the R&D system [3].

By use of this technique, we can estimate both the Young's modulus and the damping of a material. When these are combined with the Poisson's ratio and measured mass and dimensions of the beam of material, we can calculate the density, and through FEA modelling we can at last start to accurately predict the performance of a loudspeaker.

6.4 Poisson's Ratio

Most common materials such as plastics and metals become thinner in section when under tension or being stretched, the reason being that they resist a change in volume more than they resist a change in shape. This property is used later in mechanical calculations to describe exactly how various materials respond to stress and strain.

Table 6.1 shows some typical values of Poisson's ratio for some common materials.

It can be clearly seen that the vast majority of materials are in the range of 0.29 to 0.37, with only a very few materials with very different properties outside this range. Particularly relevant for loudspeakers are rubbers and honeycomb materials. Rubber materials are nearly incompressible and tend towards but *never* reach a Poisson's ratio of 0.5. Honeycomb materials are anisotropic, so their material properties can vary significantly depending upon the direction in which force and loading are applied.

Table 6.1: Poisson's Ratio Table.

Material	Poisson's Ratio
Upper limit	0.5
Rubber	0.48–<0.5
Gold	0.42
Copper	0.37–0.35
Aluminum	0.34
Polystyrene	0.34
Brass	0.33
Polystyrene foam	0.3
Stainless Steel	0.30
Steel	0.29
Tungsten	0.30–0.25
Boron	0.08
Beryllium	0.03
Re-entrant foam	−0.7
Lower limit	−1

6.5 Combined Parameters List

As we have seen in this and the previous chapter, there are many parameters. In table 6.2 we show a combined table of materials versus Young's modulus, Poisson's ratio, and damping.

Table 6.2: Combined Parameters.

Name	Young's Modulus	Density Kg/m^3	Poisson's Ratio	Damping
Aluminium (honeycomb)	15.00E+9	550	0.33	0.02
Aluminium (sheet)	75.00E+9	2700	0.33	0.01
Beryllium	250.00E+9	1800	0.33	0.01
Bextrene (p.v.a. Doped)	1.90E+9	1300	0.33	0.09
Boron	390.00E+9	2400	0.33	0.01
Boronised titanium	250.00E+9	4200	0.33	0.01
Carbon fibre composite	10.00E+9	260	0.33	0.01
Chipboard	2.00E+9	600	0.33	0.02
Copper	150.00E+9	8700	0.33	0.01
Gold	78.00E+9	19320	0.42	0.01
Graphite polymer	70.00E+9	1800	0.33	0.1
Iron	200.00E+9	7900	0.33	0.01
Magnesium	45.00E+9	1800	0.33	0.01
Paper (coated)	6.00E+9	683	0.33	0.02
Paper/phenolic	1.00E+9	350	0.33	0.06
Paper/pulp (typical)	500.00E+6	150	0.33	0.09
Plywood	8.60E+9	780	0.33	0.02
Polyamide film	3.00E+9	1400	0.33	0.02
Polyester film	14.00E+9	700	0.33	0.02
Polyethylene	1.00E+9	940	0.33	0.09
Polymethyl pentene	2.80E+9	8400	0.33	0.1
Polystyrene (foam, alloy skinned)	2.00E+9	27	0.33	0.08
Polystyrene composite	3.00E+9	950	0.33	0.02
Polystyrene foam	1.90E+6	10	0.33	0.1
PP (filled, talc)	3.00E+9	1300	0.33	0.09
PP copolymer	2.00E+9	910	0.33	0.09
PP homopolymer	1.10E+9	1000	0.33	0.09
Resin glass fibre (honeycomb)	116.00E+9	430	0.33	0.06
Rubber	2.76E+6	1124.4	0.48	0.02
Silk	6.00E+9	1000	0.33	0.01
Spider (typical)	1.00E+9	450	0.33	0.8
Titanium	116.00E+9	4500	0.33	0.01
Tungsten	4.11E+11	19300	0.27	0.01

Note

1. Named after Robert Hooke [5].

References

[1] G. Erdoğan and F. Bayraktar. "Measurement of Dynamic Properties of Materials". In: *Internoise*. Aug. 2003. `www.menet.umn.edu/%5C~gurkan/Inter%5C%20Noise.pdf` (visited on 04/02/2018).

[2] *Hooke's Law.* `https://en.wikipedia.org/wiki/Hooke%5C%27s_law` (visited on 04/02/2018).

[3] *Klippel R&D System.* `www.klippel.de/products/rd-system.html` (visited on 04/02/2018).

[4] *Material Parameter Measurement (MPM).* `www.klippel.de/products/rd-system/` `www.autodesk.com/productsmodules/mpm-material-parameter-measurement.html` (visited on 04/02/2018).

[5] *Robert Hooke: 17th century English natural philosopher, architect and polymath.* `https://en.wikipedia.org/wiki/Robert_Hooke` (visited on 04/02/2018).

PART III

Loudspeaker Models

CHAPTER 7

Small Signal Model

7.1 The Perfect Loudspeaker

This chapter is about small signal models, implying that these models are only concerned with behaviour when the loudspeaker is working at a low level. Linear conditions are also only concerned about the low-frequency region, from D.C. up until the speaker starts to depart from a mass controlled system. Quite simply these models are *not* designed to work at higher frequencies where the diaphragm affects the response.

We will start by modelling an idealised loudspeaker from a pure voltage source feeding into a perfect speaker with a known flat response. This transforms the voltage into sound correctly as shown in figure 7.1.

This forms a straight line, and if you know anything about loudspeakers, you'll know just how rare a sight such a frequency response is. Indeed, it is almost unknown.

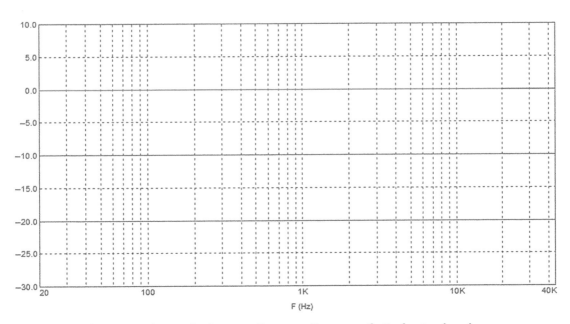

Figure 7.1: The Amplitude Versus Frequency Response of a Perfect Loudspeaker.

7.2 Low Frequency Roll-Off

For a start, conventional loudspeakers do not have a response going down to the very lowest frequencies; at best they are closer to the response shown in figure 7.2. This does not go very low, with −3 dB at 200 Hz. (This is known as the cut-off frequency, and is often strongly related to the resonance frequency.)

7.3 Small Signal Model (LR2)

Let us first look at a small signal model. This is shown in figure 7.3.

Throughout this book we are going to base these models on the LR2 model developed by Marshall Leach [1]; there are others, but in my mind this is the most practical one to use as it is reasonably accurate, and most importantly it can be modelled with real components rather than mathematical abstractions.

We will go through this model from left to right (not the usual way, we know, but hopefully things will become clearer). The first thing to notice is that the voltage is fed straight into a resistor R_e (the voice coil's resistance), so we're

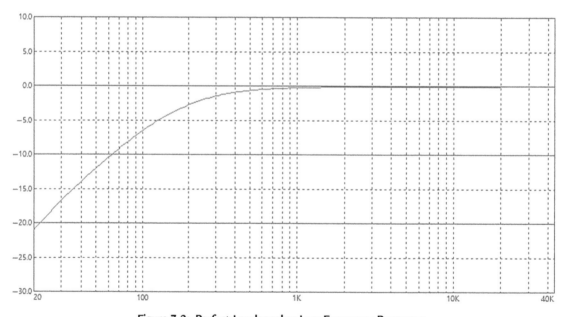

Figure 7.2: Perfect Loudspeaker Low Frequency Response.

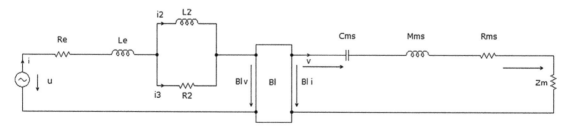

Figure 7.3: Small Signal Electrical Equivalent Circuit.

changing from a voltage controlled system to one *strongly* dependent upon current. This is confirmed later in the circuit when we see the terms *Blv* being transformed into *Bli*. This is shown as figure 7.4.

Our next element is the voice coil's inductance, to which the current (*i*) is applied to L_e. This is shown as figure 7.5, which shows a roll-off at higher frequencies.

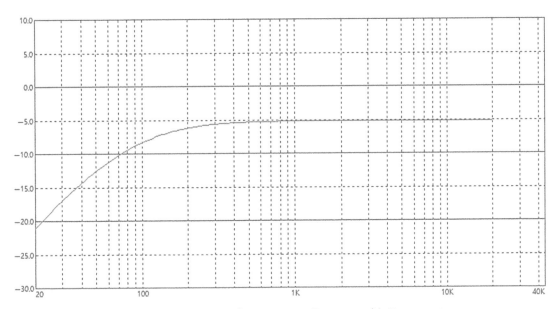

Figure 7.4: Perfect Frequency Response with R_e.

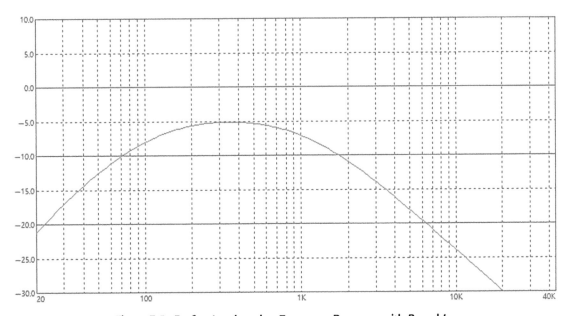

Figure 7.5: Perfect Loudspeaker Frequency Response with R_e and L_e.

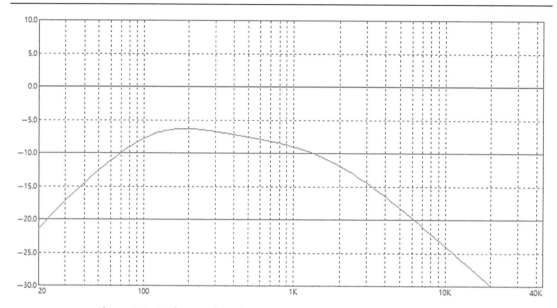

Figure 7.6: Perfect Loudspeaker Frequency Response with R_e, L_e and LR2.

A parallel pair of components is made of an inductor and a resistor, L2 and R2 (together with L_e, this threesome models the non-inductive impedance rise), the response of which is shown in figure 7.6.

Just these few parameters have had dramatic effects on the frequency response achievable from a loudspeaker, and they must be taken into account to get the desired response.

This small signal model is *only* concerned with the low-frequency region where the response is considered to be pistonic (so the whole diapragm is moving as one). This breaks down at higher frequencies and different models are required to work at frequencies where the diapragm does not move as a whole, as we will see later.

Next our model has a block[1] (Bl) that converts the signal from the electrical domain into the mechanical domain. Then on through C_{ms} and M_{ms}, which together with R_{ms} create a damped resonant network. The real effects of these components is mostly seen next when we look at the overall impedance versus frequency response. Finally we include Z_m, which represents our acoustic load.

7.4 Small Signal Parameters

Taken together these define many of our small signal lumped parameters:

- R_e is the D.C. resistance of the voice voil (Ω).
    - L_e represents the voice coil's pure inductance (H).
    - L_2 represents the voice coil's 'lossy' inductance (H).
    - R_2 represents the lossy resistance (Ω).
- Bl is the product of the flux density B and the length of active wire l (Tm).
- C_{ms} or the mechanical compliance[2] (m/N).
- M_{ms} or the moving mass (kg).
- R_{ms} or the mechanical resistance (N $\cdot$ s/m^5).
- Z_m represents our acoustic load or output.

7.5 Impedance Simulation

Figure 7.7 shows an equivalent series–parallel circuit that can be used to simulate a loudspeaker's impedance accurately.

We can clearly see here the resonant behaviour at low frequencies, together with a rising but not purely inductive impedance rise at higher frequencies. It is essential to take such factors into account, as they modify the driving force versus frequency provided by *Bl*.

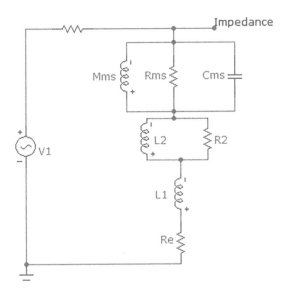

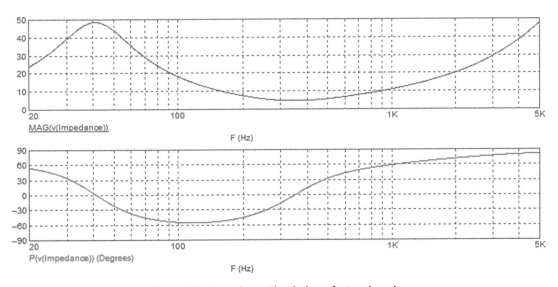

Figure 7.7: Impedance Simulation of a Loudspeaker.

We will see how these affect a loudspeaker's overall response in later chapters when we include S_d, the effect cone area (m²), which includes 1/2 of the suspension,[3] and F_s, the *free air resonance frequency* (Hz).

The S_d or effective area is a required input for calculating C_{as}, whilst F_s, the free air resonance frequency, can easily be calculated from M_{ms} and C_{ms}.

Notes

1. Known as a gyrator, it converts the serial circuit shown to an equivalent parallel one at the input.
2. 'Compliant' as in how easily something moves or changes.
3. There is some argument here as some people prefer 1/3 of the suspension.

Reference

[1]   W. Marshall Leach Jr. "Loudspeaker Voice-Coil Inductance Losses: Circuit Models, Parameter Estimation, and Effect on Frequency Response". In: *J. Audio Eng. Soc* 50.6 (2002), pp. 442–450. www.aes.org/e-lib/browse.cfm?elib=11074.

CHAPTER 8

Polynomial Models

Although polynomial models are relatively old techniques, these relatively simple underlying equations can nevertheless model much of a real loudspeaker's actual frequency response adequately, so we need to be aware of the factors involved. Let us therefore look briefly again at a perfect response, as shown in figure 8.1.

The key point here is the low-frequency roll-off. Neville Thiele proved that most loudspeakers can be modelled as high-pass filters, with a known system response function $G(s)$ built from functions of T_s, T_B, T_c, and so on [5]:

(a) For a closed system, $G(s) = s^2 T_c^2 / s^2 + s T_c^2 / Q_{T_c} + 1$

where $s = a + j(w)$ is the complex frequency variable.[1]

So, after quite a lot of algebraic manipulation, this equation can result in an amplitude versus frequency curve for the low-frequency high-pass region.

(b) For a vented or ported system, $G(s) = s^4 T_B^2 T_s^2 / s^4 T_B^2 T_S^2 + s^3 (T_B^2 T_S / Q_T + T_B T_S^2 / Q_L) + s^2 [(\alpha + 1) T_B^2 + T_B T_S / Q_L Q_T + T_S^2] + s(T_B / Q_L + T_S / Q_T) + 1$

Similar equations are available for more complex systems. We have solved these equations by hand up to sixth and seventh order bandpass systems using similar equations derived by Earl Geddes [1]. However, that was many

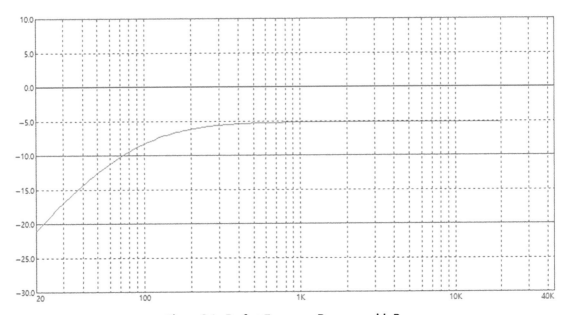

Figure 8.1: Perfect Frequency Response with R_e.

years ago, before programs were available to do this work easily. If we had to solve this today, we would probably use MATLAB [2] or SciLab [3] or another mathematics package.

Regardless of the detailed derivations, we can see an underlying structure to these polynomial equations:

- The closed system is a second order polynomial so it just has an s^2 term and an s term.
- The vented system is a fourth order polynomial with s^4, s^3, s^2, and s terms.
- A fifth order vented system with, say, an extra capacitor acting as a crossover would be a fifth order polynomial with s^5, s^4, s^3, s^2, and s terms.
- And so on, with increasing complexity of the governing equations.

Solving a seventh or eighth order system by hand is certainly possible, but the mathematics can be extremely involved. In all cases, $s = a + j(w)$ represents the complex frequency variable [4].

Because the detailed derivation and manipulation of such equations have been covered in detail in the AES journals and elsewhere, and by many people, we will not cover them or detailed derivations further. Instead, we will use programs or spreadsheets to deal directly with the parameters of the loudspeakers and the volumes of the enclosures, ports, and so forth.

However, please notice that these models are simple polynomial models. They are only concerned with the low-frequency pistonic frequency region. Specifically, they are not concerned with high-frequency regions, or any break-up when things cease to act pistonically. To deal with these problems we will need to use different tools, such as FEA, which will be covered in later chapters.

Why are we going into so much detail about these polynomial models if we are going to use other tools? The polynomial equations (or similar) underpin the low-frequency performance of all the conventional loudspeaker designs that we are concerned with, so it's essential to have a thorough understanding of what this means in performance terms, if not in the mathematical detail.

It is often easier to see this overall performance quickly using a simple model than by running a more complex FEA/BEM model, as at least initially these can have errors that prevent them working and giving you the answers you need.

Note

1. Ivo Mateljan commented that it should be: $G(s) = s^2 T_c^2 / 1 + s T_c / Q_{T_c} + s^2 T_c^2$.

References

[1]   Earl R. Geddes. "An Introduction to Band-Pass Loudspeaker Systems". In: *J. Audio Eng. Soc* 37.5 (1989), pp. 308–342. www.aes.org/e-lib/browse.cfm?elib=6090.
[2]   *MATLAB – Mathworks*. www.mathworks.com/ (visited on 05/02/2018).
[3]   *SciLab*. www.scilab.org/ (visited on 05/02/2018).
[4]   Kenneth Arthur Stroud and Dexter J. Booth. *Engineering mathematics*. Palgrave Macmillan, 2013.
[5]   Neville Thiele. "Loudspeakers in Vented Boxes: Part 1". In: *J. Audio Eng. Soc* 19.5 (1971), pp. 382–392. www.aes.org/e-lib/browse.cfm?elib=2173.

CHAPTER 9

Thiele/Small Parameters

9.1 Thiele/Small Parameters

In this chapter we'll explore three of Neville Thiele's commonly used alignments and examine certain aspects of their performance. We then will then demonstrate how these are changed into electromechanical parameters that can be directly manipulated.

It was only after these had come about that loudspeaker design began to be formalised around the system performance characteristics. From this point, there was a clear break from *bigger magnet = louder/more sensitive* towards 'targeted performance' requirement(s). This comprehensive system is now known as the Thiele/Small parameters.

Thiele's 1961 paper was subsequently republished by the Audio Engineering Society [6]. It laid down one of the first systematic descriptions of how to guarantee the fundamental performance criteria in the bass region for vented systems. It does this by using electronic filter theory and relating the low-frequency performance to Butterworth, Bessel, and Chebyshev alignments of the fourth, fifth, and sixth orders. This applies to nearly all loudspeakers (there are exceptions, but we will not deal with them here) because many loudspeakers at low frequencies can be modelled as high-pass filters of various types.

By the use of these parameters, we can describe the roll-off slope(s) as well as the shape and ripple of the responses of a loudspeaker system in terms of the polynomial models we saw previously, as well as pole/zeros, lumped parameters, gyrators, and other wonderful analogies that we saw in the chapter on small signal models.

Why did Neville Thiele feel that this was necessary? Good question. He addressed it in a lecture he made at the Audio Engineering Society in London [5].

During Thiele's early work at EMI Australia, and we are talking more than 60 years ago, loudspeaker bass response was poor.[1] Moreover, it seemed that no reliable prediction of loudspeaker performance was possible.

This was until he came up with the idea of using system alignment tables borrowed from the world of radio telemetry. In electronics, heavy use was made of filters and filter coefficients to describe their response shape(s) and roll-off rates. Thiele's genius came in applying the technique to the low-frequency performance of loudspeakers.

In his lecture he likened the overall technique as similar to using a recipe in cooking: If you follow a (good) recipe you will get an acceptable result, even without a detailed knowledge of exactly what was done.

9.2 Three Commonly Used Alignments

Let's start with a cut-down version based upon Thiele's original Table 9.1: Summary of Loudspeaker Alignments, as shown in figure 9.1.

Table 9.1: Section from Thiele's Alignment Table.

No.	Type	k	Ripple (dB)	f_3/f_s	f_3/f_b	C_{as}/C_{ab}	Q_t
3	QB3	–	–	1.770	1.250	4.460	0.259
5	B4	1.0	–	1.000	1.000	1.414	0.383
8	C4	–	0.9	0.641	0.847	0.559	0.518

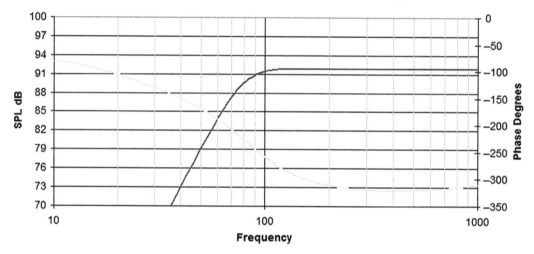

Figure 9.1: QB3 Alignment.

As you can see, we have only shown a few of the more commonly used alignments, these work on the basis of the relationships between key box and loudspeaker drive unit parameters:

- f_3/f_s or the –3 dB frequency divided by the free air resonance frequency.
- f_3/f_b or –3 dB frequency divided by the loudspeaker and box resonant frequency.
- C_{as}/C_{ab} or the acoustic compliance of the loudspeaker drive unit divided by the acoustic compliance of the box.
- Alternatively, one can calculate the ratio of V_{as}/V_{ab} or the volume of acoustic suspension divided by the volume of the acoustic box as this is an exact equivalent.
- Q_t.

Thiele's original table consists of 28 alignments, however we will only look at three of these: 3, 5, and 8.

Firstly, No. 3. Thiele describes this as a QB3 alignment as shown in figure 9.1.

This has a very safe response curve. It has relatively poor low-frequency extension but has a very smooth and ripple-free response curve with a fairly small box volume.

In No. 5, Thiele describes a B4 alignment as shown in figure 9.2. This is the 'classic' box alignment where $f_3, f_b,$ and f_s are all equal with $C_{as}/C_{ab} = 1.414$.

In No. 8, Thiele describes as a C4 alignment as shown in figure 9.3. This gives good extension with a larger box albeit with a 0.9 dB ripple in the response.

These three alignments are shown below as calculated by BoxCalc, using the same driver. From these, we can clearly see that to get good bass from this alignment we need a larger enclosure volume; also that changes to the damping or Q_t are critical in controlling response ripple.

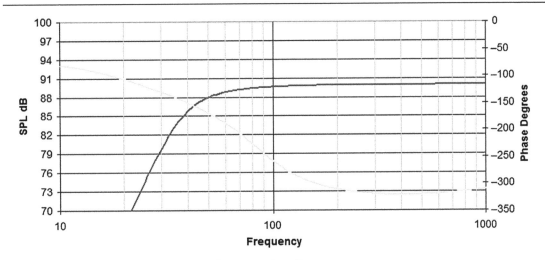

Figure 9.2: B4 Alignment.

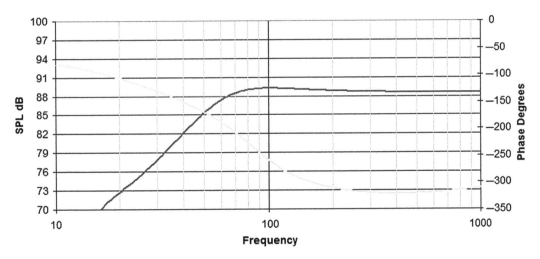

Figure 9.3: C4 Alignment.

This has serious implications for loudspeaker design at high power levels, as the D.C. resistance depends on the voice coil temperature. For a copper voice coil, the temperature coefficient is 3.9% per degrees centigrade.

So if the temperature rises 100°C the resistance will increase nearly 40%. This has two serious effects:

(i)   Less power will be made available for the loudspeaker.
(ii)   The damping will decrease or the Q_t will increase by a similar proportion.

The study of this area is covered more fully in the chapters on large signal performance; the key point to take from this is that temperature has dramatic effects on loudspeaker performance in the real world.

Also, as most loudspeakers are horrendously inefficient, significant power and therefore heat is generated producing even relatively low sound pressure levels. This applies to all moving-coil loudspeakers in a mobile phone or a high-power subwoofer. (A mobile phone does not have high powers applied, but neither can it dissipate heat as easily, and the fine wires comprising the voice coil will heat up very quickly.)

Richard Small extended Neville Thiele's work and laid down a common set of descriptors for various parameters in the acoustic, electrical, and mechanical domains, for closed box, vented, and passive radiator systems, as well as drive units for horns. Small's work was much more widely disseminated, so they are known collectively as the Thiele/Small parameters.

The following are accepted as the Thiele/Small driver parameters:

1. Driver free air resonant frequency or F_s (Hz).
2. Compliance of the acoustic suspension or C_{as}.
3. Driver electrical ratio of resistance to reactance Q_{es}.
4. Driver mechanical ratio of resistance to reactance Q_{ms}.
5. Total driver ratio resistance to reactance Q_{ts}.

We have

$$C_{as} = \frac{V_{as}}{140000} \tag{9.1}$$

and also

$$C_{as} = C_{ms} \cdot S_d^2. \tag{9.2}$$

9.3 Should You Normalise or Not Normalise?

Although using the Thiele/Small parameters makes aligning a driver and cabinet much more predictable than without using them, we have always found the reliance on 'system Qs', using parameters like V_{as} and C_{as}, makes it difficult to relate to the directly controllable underlying parameters.

As an example, let us take as starting points Thiele and Small's work on the normalised amplitude versus frequency response of the closed box system for different values of total system $Q_t = 0.5, 0.707, 1.0,$ and 1.5. We have plotted these using SpeakerPro as shown in figure 9.4.

Then let us instead take the same data and plot it as real SPL values using BoxCalc^2 as figure 9.5.

These look completely different, don't they? In normalising the curves, as Thiele and Small both do, there is a danger of losing sight of the other changes in both sensitivity and around the roll-off that the different Q_t values imply. In practice, a higher Q_t means a weaker electromagnetic system and vice versa.

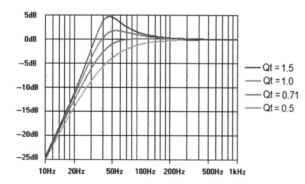

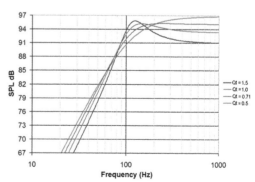

Figure 9.4: $Q_t = 0.5, 0.707, 1.0,$ and 1.5 Normalised.   **Figure 9.5:** $Q_t = 0.5, 0.707, 1.0,$ and 1.5.

As far as we are concerned, the big problem of relying on the Thiele/Small parameters and alignment methods is that these parameters are not exactly easy to manipulate. If you are really keen, read up the theory from Thiele and Small republished by the Audio Engineering Society[3] along with many other important papers of the time.

The first paper that we are aware of that tackles this situation head on is 'Theoretical and Practical Aspects of Loudspeaker Bass Unit Design' by Garner & Jackson [1]. Here a systematic approach is developed that uses the Thiele/Small parameters to define the required response shape.

The paper then goes straight into converting from this desired response shape into realisable electromechanical parameters. We have extended this work into programs that allow one to swap between the Thiele/Small parameters and the electromechanical parameters, and also a spreadsheet that allows one to make a choice between different alignments based on the enclosure size and chosen sensitivity; the spreadsheet then shows the electromechanical parameters for this choice.

Having said all that, there are a couple of problems with the original Thiele/Small parameters. Firstly, if applied to drive unit design, they do not fully incorporate damping; and secondly, they are rather fragmented when used with different types of systems. These concerns were addressed by Ernest Benson, who proposed a unified model that is used in computer modelling to this day.

Earl Geddes has made a thorough study of bandpass systems of fourth to eighth order in his paper 'An Introduction to Band-Pass Loudspeaker Systems' [2]. All of these include polynomial models. Don Keele applied Thiele/Small parameters to low-frequency horn design [4].

9.4 Electromechanical Parameters

Obviously, these are all real, measurable quantities, and most importantly they are all easily calculated from the original Thiele/Small parameters. Before we had the Thiele/Small parameters, they were all we had, and now with FEA modelling all are directly available:

- R_e is the D.C. resistance of the voice coil (Ω).
- Bl is the product of the flux density B and the length l of wire in the gap (Tm).
- C_{ms} or the mechanical compliance (m/N).
- K_{ms} is the inverse of the mechanical compliance (N/m).
- M_{ms} or the moving mass (kg).
- R_{ms} or the mechanical resistance (N $\cdot s/m^5$).
- S_d is the effect cone area (m^2).
- F_s is the free air resonance frequency (Hz).

This leaves only damping as a dimensionless quantity. This is less amenable to direct control, though if you can work out the mechanical resistance (R_{ms}), this becomes simple as well. Ultimately, these are the only parameters that you have direct individual control over; it is in manipulating these that one can exercise real design choices.

We have extended this understanding into programs that allow one to swap directly between the Thiele/Small parameters and the electromechanical parameters. Also, a spreadsheet format allows direct choice of practical alignments and outputs, along with the main electro-mechanical parameters, which are under direct control. These are available from [3].

Many measurement systems now exist that put out parameters in both sets. By their very nature, these are fully interchangeable (or even both may be used simultaneously), thus giving the designer the freedom to work in whichever domain suits them best.

As a designer it is essential to work with real quantities that can be measured and checked. This means using the electromechanical parameters, but at the same time, the Thiele/Small parameters do allow the overall relationships between these and the cabinet to be seen in a clearer light.

However, is all well? We would argue that it is not. Just because it is easy to see how a driver will align with a cabinet does not mean you are going to get consistent or desirable results.

To do this needs a fundamental understanding of the parameters involved and how to change them. It is here that we feel that the modern approach of much software partly fails because it cannot encompass the multitude of potential alignments in a meaningful manner. It can show a curve, but it tells you nothing about what that bass alignment will sound like. And do not assume that just because you have settled on a specific bass alignment, the loudspeaker as a whole will give a satisfactory performance.

For a start, you need to define what a satisfactory performance for your application is. For a hi-fi loudspeaker, for example, amongst the important criteria is a general acceptance that an even response over a reasonably wide range of angles is desirable.

More specifically, for our subwoofer we need to ensure that it is capable of the dynamic range and output that we will require, so the standard electromechanical parameters will need extending to large signal[4] as well as small signal conditions.

Notes

1. Some would argue that things have still not changed for the better.
2. One of the tools available at Geoff Hill's website [3].
3. See *Loudspeakers Vol.1* in references.
4. See *IEC Standard IEC62458 Sound System Equipment—Electro-Acoustic Transducers—Measurement of Large Signal Parameters* for details.

References

[1]   A. V. Garner and P. M. Jackson. "Theoretical and Practical Aspects of Loudspeaker Bass Unit Design". In: *Audio Engineering Society Convention 50*. Mar. 1975. www.aes.org/e-lib/browse.cfm?elib=2485.

[2]   Earl R. Geddes. "An Introduction to Band-Pass Loudspeaker Systems". In: *J. Audio Eng. Soc* 37.5 (1989), pp. 308–342. www.aes.org/e-lib/browse.cfm?elib=6090.

[3]   *Geoff Hill website*. www.geoff-hill.com/ (visited on 02/02/2018).

[4]   D. B. (Don) Keele Jr. "Low-Frequency Horn Design Using Thiele/Small Driver Parameters". In: *Audio Engineering Society Convention 57*. May 1977. www.aes.org/e-lib/browse.cfm?elib=3105.

[5]   *Neville Thiele AES London 2009*. www.aes-media.org/sections/uk/meetings/AESUK_lecture_0911a.mp3 (visited on 05/02/2018).

[6]   Neville Thiele. "Loudspeakers in Vented Boxes: Part 1". In: *J. Audio Eng. Soc* 19.5 (1971), pp. 382–392. www.aes.org/e-lib/browse.cfm?elib=2173.

CHAPTER 10

Large Signal Domain and Model

Before starting the real design work, the large signal domain and modelling need to be discussed, as these techniques help us to understand what happens when we use a loudspeaker with real-world signal levels. Previous analysis has all been in the small signal domain, but in the large signal domain this breaks down, so we need different techniques. In the small signal domain it was not the absolute signal level that mattered but rather that the signals used were sufficiently small to ensure that linear relationships held true.

We saw in the Thiele/Small analysis that most conventional moving coil loudspeakers are horrendously inefficient. Unfortunately, efficiency has dropped significantly over the past 50 years. When Thiele wrote his paper in 1961, the typical efficiency for a medium sized loudspeaker was around 1–4%.

Thiele commented that some of the alignments he showed could be considered suspect as they required a boost of four times at resonance and that this would push into areas where his underlying assumptions of low efficiency would break down.

Fast forward to 2018 and the typical loudspeaker in a mobile phone is often around 0.01% efficient. So now there is no chance of that problem; however, advances in DSP now allow us to accurately make such corrections available with effectively zero cost. It is also true that the loudspeakers today are much smaller, and have nowhere near the bass response, even to the point that the bass response of mobile devices has been redefined to mean the response below 1 kHz.

We make this point because it's not actually the frequency range that matters but rather the displacement/excursion as compared to the size and depth of the loudspeaker. Many loudspeakers used in mobile devices are required to make huge displacements or excursions relative to their size. So in the large signal domain, we are interested in any changes that happen to a loudspeaker when it is operated outside of the small signal domain where, at least by implication, the behaviours are linearly predictable.

We have many limited or nonlinear excursion issues, these can show up as non-symmetrical $Bl(x)$, $C_{ms}(x)$, and $L_e(x)$ curves, which individually or combined can cause either physical limiting or many types of distortion.

The Klippel LSI [3] approach builds a mathematical model of the particular loudspeaker while it is being exercised over a controlled displacement/excursion, distortion, and temperature or power range. It then produces graphical outputs of the measured variations of an actual physical loudspeaker together with first to eighth order coefficients, which describe these curves, along with distortion predictions for each of these separate mechanisms.

The lower and lower efficiency of modern loudspeakers requires even higher and higher power to be used; the low efficiency means that the majority of this power eventually ends up as heat. This heat causes many problems. Apart from changing damping Q_{es}, in design, we should and can reverse this process by simulating the required parameters over a desired excursion or power range, thus ultimately predicting performance at high power levels or excursion.

A general rule for conventional moving coil loudspeakers is that for any given sound pressure level, provided the physical size is significantly smaller than the wavelength of the frequency being produced, a smaller loudspeaker will be less efficient and have to move further (have greater excursion) than a larger loudspeaker.

So in order to handle all these variables, we need to extend our model from the small signal domain into the large signal domain.

We will see that the small signal model has been changed by adding (x), (T_v), and sometimes (i) after various key parameters that are subject to change.

We suggest that the key parameters to concentrate on are as follows:

- $R_e(T_v)$.
- Bl, which becomes $Bl(x)$.
- C_{ms}, which becomes $C_{ms}(x)$ although its inverse $K_{ms}(x)$ is easier to visualise.
- L_e, which becomes $L_e(x)$.

Measuring these variables is a subject in its own right and we would recommend the papers on this subject published in the AES by David Clarke [1], Wolfgang Klippel [2], and many others.

Fortunately, we are not concerned with measuring the parameters (though that would be nice) but rather we should like to predict these parameters whilst we are still at the design stage, as here improvements can be made with little or no extra cost.

$R_e(T_v)$—this is primarily down to keeping the voice coil temperature down and stopping any fluctuations in temperature, which can be tricky, so air flow and thermal mass have a complex role to play here which we will not go into in detail.

$Bl(x)$—fortunately, this one is relatively easy to model and later, in the chapter on motor unit design and the appendix on FEMM, we will see this approach being used in predicting the $Bl(x)$ curve.

We have used this technique for over 10 years now and have found that, when correctly modelled, it gives results within ± 5 per cent of a physical motor unit as confirmed by the Klippel LSI module. This is good, as nonlinearities and non-symmetries in $Bl(x)$ show up clearly as predictable distortions.

$C_{ms}(x)$ (or its inverse $K_{ms}(x)$) gets its own chapter. Here we will use mechanical FEA to predict the force deflection curve(s). This is important, as lack of control here can cause problems ranging from bottoming and popping through to various types of distortion.

$L_e(x)$—fortunately, most of the distortions from $L_e(x)$ much lower in level and generally more benign than those of $Bl(x)$ and $C_{ms}(x)$. Also, these distortions from $L_e(x)$ are dependent on the current flowing as well as the displacement, so it requires quite complex modelling.

$R_{ms}(x)$ stands for the mechanical resistance which varies with x, the excursion or displacement.

$F_m(x, i, i_2)$ is the reluctance force from the voice coil and magnet and varies with x (the excursion or displacement) and the currents, i and i_2.

Z_m represents the output as shown in the electromechanical domain rather than as an acoustic output.

Please note that in theory, an extension of FEA should be able to calculate the impedance versus frequency.[1] However, the analytical theory[2] of a loudspeaker's impedance is still a point of discussion; FEA should be able to take account of effects such as skin depth versus frequency.

Thus, we should be able to accurately predict the impedance response. However, we will not be modelling effects that include $L_e(x, i)$, $L_2(x, i)$, and $R_2(x, i)$ at this time. Another area of significant influence is that of power compression. This is usually related to temperature, especially that of the voice coil, but if a loudspeaker is driven hard enough for long enough, the whole structure can and does heat up. In that case, the thermal mass of the rest of the motor unit can store a lot of heat that will heat up the voice coil even during intervals of very low signal; thus this can directly affect how the loudspeaker performs when it is cold, warm, or hot! Such temperature variations have the power to influence pretty well any of the mechanical elements, especially those where stiffness is involved, as this often has a direct temperature coefficient.

10.1 Large Signal Model

Earlier we looked at the small signal domain and we defined many of our small signal parameters. I will repeat them here, as we will use them again, albeit mostly in slightly different form. See figure 10.1.

- R_e is the D.C. resistance of the voice coil (Ω).
  - L_e represents the voice coil's pure inductance (H).
  - L_2 represents the voice coil's 'lossy' inductance (H).
  - R_2 represents the lossy resistance (Ω).
- Bl is the product of the flux density B and the length l of active wire (Tm).
- C_{ms} or the mechanical compliance (m/N).
- M_{ms} or the moving mass (kg).
- R_{ms} or the mechanical resistance (N $\cdot s/m^5$).
- Z_m represents our acoustic load or output.

We will now look at our large signal model:[3]

At first sight this looks remarkably similar to—and indeed it was developed from—the small signal model. However, here we have various additions to the subscripts; the components are highlighted in red (figure 10.1).

These, Klippel describes as the 'state variables', and these things are subject to change due to excursion (x), current (i), or temperature (T_v)[4]. Also, note C_{ms} has been replaced by $1/K_{ms}$, its inverse. This is because it is easier to visualise its effects as K_{ms}. We have one extra parameter, $F_m(x, i, i_2)$.[5]

So, our large signal parameters now become:

- $R_e(T_v)$.
- $L_e(x, i)$.
- $L_2(x, i)$ and $R_2(x, i)$.
- $Bl(x)$.
- $C_{ms}(x)$ but in preference we will use its inverse, $1/K_{ms}(x)$.
- M_{ms}.
- R_{ms}.
- $F_m(x, i, i_2)$.
- Z_m.

Again as in the small signal model, we add:

- S_d, the effective diaphragm area (m^2).
- F_s, the free air resonance Frequency (Hz).

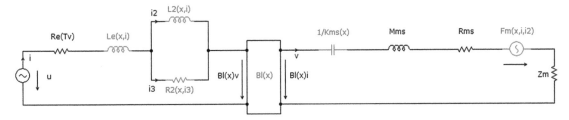

Figure 10.1: Large Signal Electrical Equivalent Circuit.

The S_d or the effective area is a required input for calculating C_{as}. We will cover this in detail in Chapter 16: Driver Design, on page 71, whilst F_s or the free air resonance frequency (Hz) can easily be calculated from M_{ms} and C_{ms}, as we saw earlier.

However, the aim of this chapter was to introduce you to these various terms so you understand what they represent. Later on we will see how they affect a design.

Notes

1. With appropriate meshing to assure the skin depth is correctly modelled.
2. As of 2017.
3. Developed by Wolfgang Klippel.
4. Strictly speaking, the input (u) is a state variable as well, but we are or will be designing with this in mind.
5. This is after the gyrator, as it is a reluctance force involving inductance and its interaction with the magnet. It is dependent upon the rate of change of current and inductance versus excursion (x).

References

[1]   David Clark. "Precision Measurement of Loudspeaker Parameters". In: *Audio Engineering Society Convention 99*. Oct. 1995. www.aes.org/e-lib/browse.cfm?elib=7685.

[2]   Wolfgang Klippel. "Dynamical Measurement of Non-Linear Parameters of Electrodynamical Loudspeakers and their Interpretation". In: *Audio Engineering Society Convention 88*. Mar. 1990. www.aes.org/e-lib/browse.cfm?elib=5791.

[3]   Wolfgang Klippel. "Large Signal Performance of Tweeters, Micro Speakers and Horn Drivers". In: *Audio Engineering Society Convention 118*. May 2005. www.aes.org/e-lib/browse.cfm?elib=13138.

PART IV

The Design Process

CHAPTER 11

There Is a Job to Do—But Exactly What?

In this chapter, we start to consider loudspeaker design in the widest sense. Taking the example of a subwoofer, we begin to break it down into bite-sized chunks.

Where should we start? We always try to start from what final result is needed, by asking the question: 'What are we trying to do?' Then we can ask and answer questions about the look and feel of a loudspeaker.[1]

So:

- Do you want to build a home cinema system?
- A loudspeaker for your kitchen or bathroom?
- A subwoofer for your car?
- The loudspeaker part of a conventional hi-fi or home theatre system?
- Something to improve the sound from a tablet or smartphone?
- A public address system?
- A loudspeaker to reproduce a bass guitar?
- A miniature or micro loudspeaker for a tablet, smartphone, or multimedia device?
- Will it need to reproduce the full range?
- Will it just need to reproduce high frequencies?

Each of these will require changes in the detail as well as the overall design. All of these can be designed using conventional moving-coil loudspeakers.

Note

1. Also known as the industrial design or ID.

CHAPTER 12

Common Questions for Most Loudspeakers

Herewith some answers to the more common questions needed to break a loudspeaker design down into smaller component pieces. These can then be looked at either individually or collectively to ensure we are aiming at the correct specification(s).

An engineer may be called on to design for many practical applications, all of which may have different requirements. However, we should start off with a set of standard parameters and questions that we believe must be answered:

1. What maximum sound pressure level (SPL) is required, under what conditions?
2. What power is available from the amplifier or sensitivity is required?
3. What are the required −3 dB frequencies—low and high frequencies?
4. What cabinet or enclosure volume is available?
5. What type of loudspeaker system is required?
6. What size and how many loudspeaker(s) are being envisaged?

That's just six key questions, and in our view you need to answer them all before proceeding further. Now it is time for a sanity check. Ask yourself if the answers are actually realistic.

As an example a single 1 inch or 25 mm metal dome tweeter would probably *not* be able to meet every item in the following idealised specifications:

- Max SPL equals 120 dB SPL at 1 m.
- Maximum Power equals 100 W (RMS, 20 Hz to 50 kHz).
- −3 dB at 20 Hz and at 100 kHz.
- Operates in free air without further acoustic loading.

Deciding whether these are really feasible or not should act as a sanity check. It is only too easy for someone with only a passing familiarity to write a seemingly sensible specification which is anything but practical. Only when you have decided it is feasible should you answer the following questions:

1. Are there any particular off-axis requirements?
2. Are there any particular distortion requirements?
3. What are the frequency response requirements?
4. Are there any environmental or safety factors?
5. Are there any physical constraints?
6. Are there any visual constraints or requirements?
7. What are the cost or equipment limitations?
8. Are performance tolerances important between drivers or systems?

These are just some of the questions that need to be answered before design can start. A big problem that most designers face is the lack of answers to these questions. Either that or there are often assumptions (usually unsaid ones) made about these.

The tolerances issue is a particular bone of contention in the industry, probably causing more grief than nearly any other part of the design and manufacturing process. As we were reminded recently, these tolerances can and do interact. Such tolerances need to be considered, as nothing can be exact.

'What is an acceptable variation?' A particular problem with the modelling process is the underlying assumption that everything can and will be made the same. However, in the real world and when using actual parts this is not always the case. Variations and tolerances are covered in more detail in Appendix L on page 325.

Try to remember the saying: 'Assumptions are the things that make an ass out of you and me.' To the list must be added questions about construction methods, materials, facilities, quantities, and time scales.

Being completely realistic, we need to consider whether this project is completely new, whether you can borrow skills, knowledge and facilities, or whether you need to start from scratch. For the purposes of this book we will assume the latter and go from there.

CHAPTER 13

Specifying a Loudspeaker Driver

We examine the essential underlying system specifications and what other parameters need to be included, taking into account the type of enclosure or system before considering the makeup of individual drivers and how they are being operated.

What process should we go through when designing a drive unit from scratch?

1. We need to know the underlying system specifications:
   - What maximum system sound pressure level is required?
   - What is the minimum system impedance?
   - How efficient does the system need to be?
   - What power does the system need to be able to handle?
   - Sound pressure level requirements:
     - What variation in SPL is acceptable?
     - Over what frequency range—overall trend
     - On-axis
     - Off-axis
     - Distortion
   - Do we need to consider:
     - Time alignment
     - Waveform shape—group delay
     - Other factors
   - What types of enclosure(s) make up the system?
     - Closed box
     - Vented or ported
     - Auxiliary bass radiator
     - Bandpass
     - Horn
     - Open baffle
2. Decide individual driver specifications:
   - Frequency range of driver
   - Size of enclosure driver needs to work in
   - Underlying driver sensitivity
   - Underlying driver frequency response trend
   - Low frequency $-3\,\mathrm{dB}$ Frequency (Hz)
   - High frequency $-3\,\mathrm{dB}$ Frequency (Hz)
   - Sound pressure levels required:
     - At what distance
     - Under what conditions:
       - Free space
       - Ground plane or floor
       - Floor and wall

- ■ Floor and two walls
- ■ In wall
  - ○ Power handling
3. Individual driver performance requirements:
   - • Underlying electromechanical parameters
   - • Large signal performance requirements
   - • Desired frequency response characteristics
     - ○ On-axis
     - ○ Off-axis
       - ■ Distortion
       - ■ Time alignment
       - ■ Waveform shape—group delay

CHAPTER 14

Project Planning and a Bill of Materials

We strongly recommend producing an outline plan, highlighting how this differs from a project plan and discussing why this difference is important and how it can be developed later. A simple example can then be used to form the core of such a plan.

We would suggest producing some form of project plan and doing this before you start any 'designing', especially if you think that the design, building, and testing stages are more interesting,[1] as a project plan implies a situation where you know exactly what you are doing.

But in this book we are exploring the design and modelling of loudspeakers, and there is a difference between knowing exactly what you are going to do and planning for it. Please do not worry about doing this in detail, as a project plan can be as simple as a list of questions.

We'd strongly suggest that these questions should be in the form of the key project steps, and these will certainly need to be included. We suggest we start with a spreadsheet for this, simply as it's easy to keep to a common column layout. So, the reason why should we use some form of project plan is as follows:

(i)   To help ensure you have the resources you need to complete the task(s).
(ii)   To help to make sure you have accounted for as many tasks as possible.
(iii)   To help to give you some idea of how long the project will take.
(iv)   To help highlight if you are working to a deadline or at critical point(s) in the plan.

Formal project planners include an automatic scheduler at their heart. They generally include the following core pieces of information: earliest task start date; latest start date; estimated task duration; actual task duration; task precedents; task successors; resources available; and many others. You can force a spreadsheet to do this; a quick search online should show you how.

The scheduler puts this information together and can produce a tracking Gantt chart (as well as other guides to progress). A Gantt chart (especially a tracking Gantt) allows one to see the progress or otherwise of a project at a glance. One of the most difficult elements in a project is estimating how long a task will take. Somewhat easier if you have done a task before, it can provide an overview that few other techniques can.

Just having a thought-out plan or plans will dramatically increase your chances of successful completion. It will enable you to call on help or other resources by indicating where and when you need them, or enable you to know when to back off one thing and get on with another.

Nearly every 'failed' project we have done has involved some lack of planning, faulty assumptions, or a lack of clarity about the project as a whole. A proper plan would have focused attention and helped to resolve these issues.

The essence of a project plan is an ordered list of tasks and their relationship to other tasks. Initially this part is best first put together in a spreadsheet; later one can plot Gantt charts and the rest, but let's start simply. An example plan is shown as figure 14.1.

	A	B	C	D	E	F	G	H	I	J	K	L
1		Project Schedule									HA=Hill Acoustics	
2		Range/ Model:										
3		Date: 22.02.15			Supplier: TBA				Edited By: Geoff Hill			
4		Revision: A									Company	Person
5	Process #	Task Description	Schedule	Supplier schedule	Actual Date	No of Days	Status Description	Responsible	Remarks	Responsibility	Responsible	
6	1	Initial Research Document	28/09/2015		26/09/2015	10		GH		HA	GH	Geoff
7	2	Establish Initial Brief	08/10/2015		08/10/2015	10		GH		HA	GH	Geoff
8	3	Brief Sign Off						GH		HA	GH	Geoff
9	4	Initial Sketches and foams	10/11/2015		10/11/2015			TA		HA	TA/AP	Tom
10	5	Initial Acoustic Design (see sub tasks)	10/11/2015		08/12/2015			GH		HA	GH	Geoff
11	6	Final Design Selection	10/11/2015		10/11/2015			GH/AS		HA	GH	Geoff
12	7	Complete Inventor/ 3D Model	03/12/2016		03/12/2016			TA		HA	TA	Tom
13	8	Complete Initial Drawings	13/01/2016					TA		HA	TA	Tom
14	9	Acoustic Design (see sub-tasks)	14/01/2016					GH		HA	GH	Geoff
15	10	Drawing Approvals (internal)	14/01/2016					AS		HA	AS	Alan
16	11	Initial Data/ drawings released	15/01/2016					TA		HA	TA	Tom
17	12	Initial Acoustical Integration (see sub task)	20/01/2016					GH		HA	GH	Geoff
18	13	BOM Creation	20/01/2016					GH		HA	TA	Tom
19	14	Supplier quotation submission	24/12/2015		10/01/2016			CW		Supplier x 3	CW	Colin
20	16	Business Award/ Cost approvals	04/01/2016		04/02/2016			GH		HA	GH	Geoff
21	15	Mutual Supplier and Company QC Standards to be written and agreed by both parties.	08/02/2016					AS		HA & Supplier	AS	Alan
22	17	Tooling Release	08/02/2016					GH		HA	GH	Geoff
23	18	Packaging Design	22/03/2016					AS		HA	TA	Tom
24	19	Artwork and logo design	22/03/2016					TA		HA	TA	Tom
25	20	Tooling completion	28/03/2016					CW		Supplier	CW	Colin
26	21	1st sample submission (drivers) - Hand Built Samples	11/04/2016					CW		Supplier	CW	Colin
27	22	1st off tool sample submission (mechanical parts)	11/04/2016					CW		Supplier	CW	Colin
28	23	1st sample testing and evaluation (drivers)	11/04/2016					GH/GH		HA	GH/GH	Geoff/Geoff
29	24	1st off tool sample comment (mechanical Parts)	18/04/2016					TA		HA	TA	Tom
30	25	2nd sample submission (drivers) - Line Built Samples	09/05/2016					GH/GH		Supplier	GH/GH	Geoff/Geoff
31	26	2nd Off Tool sample submission (mechanical Parts)	09/05/2016					CW		Supplier	CW	Colin
32	27	2nd Sample testing and evaluation (driver)	16/05/2016					GH/GH		HA	GH/GH	Geoff/Geoff
33	28	2nd Sample electonic/ functional performance comment	16/05/2016					GH		HA	GH	Geoff
34	29	2nd off tool sample comment (mechanical Parts)	16/05/2016					TA		HA	TA	Tom
35	30	Mechanical Parts approval	23/05/2016					TA		HA	TA	Tom
36	31	Electronic/ functional performance comment	23/05/2016					GH		HA	GH	Geoff
37	32	User manual info Submission	23/05/2016					AS		HA	AS	Alan
38	33	Voicing process commence	23/05/2016					GH		HA	GH	Geoff
39	34	Voicing process completion						GH		HA	GH	Geoff
40	35	Working sample submission (complete) #pcs required?	23/05/2016					CW		Supplier	CW	Colin
41	36	Commence reliability testing program (Leigh)	30/05/2016					AS		HA	AS	Alan
42	37	Driver approval / Select Golden Sample Drivers, produce Limits and Send to Company, Supplier	30/05/2016					AS		HA	AS	Alan
43	38	Commence Electrical Safety testing	30/05/2016					CW		Supplier/ HA?	CW	Colin
44	39	Working sample comments or approval	30/05/2016					AS		HA	AS	Alan
45	40	Tooling/performance adjustment	13/06/2016					CW		Supplier	CW	Colin
46	41	Final BOM release	25/07/2016					CW		Supplier	CW	Colin
47	42	Pilot run	25/07/2016					CW		Supplier	CW	Colin
48	43	Complete Electrical Safety testing Aprovals.	25/07/2016					CW		Supplier/ HA?	CW	Colin
49	44	PP units submission	01/08/2016					CW		Supplier	CW	Colin
50	45	PP units approval	08/08/2016					AS		HA	AS	Alan
51	46	Reliability testing complete/ issue results	08/08/2016					AS		HA	AS	Alan
52	47	Start Production	29/08/2016					CW		Supplier	CW	Colin
53	48	First shipment	12/09/2016					CW		Supplier	CW	Colin
52	47	Start Production	29/08/2016					CW		Supplier	CW	Colin
53	48	First shipment	12/09/2016					CW		Supplier	CW	Colin
54												
55					SUB TASKS							
56		5.INITIAL ACOUSTIC DESIGN										
57		Task Description	HA schedule	Supplier schedule	Actual Date		Status Description	Responsible	Remarks	Responsibility	Responsible	
58	5.1	Theoretical Parameters from Initial Brief						GH		HA	GH	Geoff
59	5.2	Produce Initial Motor Unit Design						GH		HA	GH	Geoff
60	5.1	Generate LPM Parameters						GH		HA	GH	Geoff
61	5.2	Alignments						GH		HA	GH	Geoff
62		9.ACOUSTIC DESIGN										
63		Task Description	HA schedule	Supplier schedule	Actual Date		Status Description	Responsible	Remarks	Responsibility	Responsible	
64	9.1	Produce Detailed Motor Unit and Driver Design						GH		HA	GH	Geoff
65	9.2	Output - Overall Response						GH		HA	GH	Geoff
66		12.ACOUSTIC INTEGRATION										
67		Task Description	HA schedule	Supplier schedule	Actual Date		Status Description	Responsible	Remarks	Responsibility	Responsible	
68												
69		Green background: Finished										
70		Yellow background: Slight Delay										
71		Red Background: Serious Delay										
72		U.K > production release										
73												

Figure 14.1: Example of a Project Plan.

For the next stage, before starting on design work, we strongly recommend producing a bill of materials (BOM) (or rather the core structure of one). What should be in a BOM, and how should it be structured? We prefer a hierarchy:

- Cabinet Assembly
    Trims
    Screws
    Gaskets
    Mounting Bolts and Nuts
    Input Terminals XLR/Phono—Digital?
    Mains Input Terminal Switch and Fuse
    Grille(s)
    Feet
    Packaging
      Cardboard
      Leaflet
      End Caps
- Amplifier Assembly
    Printed Circuit Board
    Capacitors
    Resistors
    Semiconductors
    Power Amplifier
    Heatsink
- Subwoofer Driver
    Pole
    Chassis
    Top Plate
    Magnet
    Voice Coil Assembly
    Chassis
    Cone and Surround Assembly
    Dust Cap
    Spider or Damper
    Connection Terminals
    Glue(s)
    Screws and Nuts
- Any other parts or processes required.

It is essential to start a BOM early. Note: it does *not* have to be complete! We find a spreadsheet ideal, though others prefer a database. Some 3D drawing packages such as Fusion 360 (dealing with assemblies and sub-assemblies) have a built-in database structure; please just do it.

The BOM approach has many uses, as it will both take from and feed into the project plan, ensuring all the necessary parts have been sourced or designed and produced. It also allows the tracking of costs and suppliers for the drivers(s), cabinet stages, and any other parts. Even doing nothing other than a bill of materials, this will be a big help, as it will enable one to focus on a component part.

So let us look at the example of a subwoofer driver parts list as figure 14.2.

We can then produce equivalent parts lists for the voice coil, the cabinet assembly, and the amplifier as required.

Project:		Loudspeaker Driver Model:			Date:		
#	Quantity	Description	Size(s)				
1	1	Backplate & Pole	110mm OD x 50mm ID x 50mm	Part No	Supplier	Cost	
2	2	Magnet	120OD x 60 ID x 20H				
3	1	Top Plate	53.00mm ID				
4	2	Suspension	50.5mm ID				
5	1	Voice Coil	4Layer 0.3mm x 50.0mm ID				
6	1	Diaphragm	50.2mm ID				
7	1	Dust Cap	50.2mm ID x 85mm OD				
8	1	Surround	250mm OD x 15mm H				
9	2	Terminal Post	4mm Terminals				
10	1	Gasket	300mm OD x 250mm ID x 1mm Thk				

Figure 14.2: Example of a Loudspeaker Driver Parts List.

Note

1. We say this because we have been known to plunge ahead on a task that seems more interesting to ourselves, to the detriment of the project as a whole.

CHAPTER 15

Designing a Subwoofer

In this chapter, we discuss some of the project stages required to transform our ideas into an actual loudspeaker—the stages we will consider here are the physical and mechanical aspects.

Our first design will be a subwoofer designed to enhance the bass performance of a home hi-fi or home cinema system. So where do we start?

- First let us pin down, even if roughly, the overall physical size;[1]
- Then the type of system.

We will use a closed box subwoofer cabinet 450 mm × 450 mm × 450 mm. We will use 25 mm thick material for this, so we are ignoring the volume taken by the loudspeaker driver for the moment. We will have 400 mm × 400 mm × 400 mm; this equals a volume of 0.064 m³, or 64 litres, as it is usually stated.

We will continue the motor unit (magnet assembly and voice coil) design later. But for now, let us concentrate on the mechanical aspects. We need to consider the physical strength of the chassis, noting that the chassis not only has to meet all of its requirements under normal operational conditions but it also needs to do so after abusive handling or drop testing.

There are many types of chassis in use. Nowadays, mostly plastic injection mouldings are used for high volume products. In the past, high-volume designs tended to be made from pressed steel; however, individual designs are sometimes made from die castings or even machined from castings or solid materials. More modern alternatives now include rapid prototyping methods such as stereolithography and 3D printing. These have been available to industry for some time and are now becoming much more readily available and affordable for the small company and even a hobbyist.

A typical requirement for a chassis is that it should be stiff and that it should allow the motor unit to transfer all of its force to the cone/surround assembly. However, it also needs to be strong and not brittle to allow it to handle reasonable physical abuse from drops or other shocks without suffering damage.

So not only do we need to consider the visual, mechanical, and acoustical aspects of all of the parts in a loudspeaker, we also need to look at how we are going to produce our speaker and its parts. In turn, we will need to consider any quantities that we need to build, as well as the tooling and facilities available.

So what aspects of the chassis do we need to consider? Well, what is the function of a chassis? Quite simply, it has two functions:

- To show off visually—we will not go into this, but a lot of loudspeakers seemed to be designed with this foremost.
- To keep all of the parts in the correct alignment at all times.

But there is a problem here: To design the chassis we need to know quite a bit about the motor unit, particularly the weight of it. Also, we must know about the cone and surround to make sure it fits together correctly.

So we need to know about the cone and surround assembly, considering the size, shape, weight, and material of the cone, dust cap, and surround, ensuring that they are capable of the linear movement required and that no breakup modes are causing significant problems.

We will deal more fully with the physical, mechanical, and acoustical performance aspects of all of these parts in their own chapters. As you can see, we are going round and round in circles here and getting nowhere!

So that is how *not* to go about designing a chassis and, most importantly, 'why'. So let us start again:

Our design will be of a closed box subwoofer cabinet 450 mm × 450 mm × 450 mm and a driver to fit into it.

- We will use 25 mm thick material 450 mm × 450 mm × 450 mm (external).
- This equals a 400 mm × 400 mm × 400 mm (internal) or a volume of 64 litres.
- Next, we would strongly suggest, we *roughly* design the cone and surround.
- Then we can produce an initial motor unit design.

From these, we will know the major sizes, weights, and mechanical dimensions and specifications required for our chassis design. Now we can either get on and do the detailed mechanical design ourselves or pass it onto someone else. Either way, we will have confidence that the core design is mechanically and acoustically viable at an early stage.

Note

1. We are not going to be drawn into discussing visual aspects.

PART V

What's Really Going On Inside a Loudspeaker?

CHAPTER 16

Driver Design

In this chapter, the underlying electromechanical parameters are drawn from an extension of the Thiele/Small parameters, using the method from Garner and Jackson's 'Theoretical and Practical Aspects of Loudspeaker Bass Unit Design' [1]. From the basic equations we can then produce the detailed specifications and feed them into an FEA model. We also discuss the type of FEA model that's most appropriate.

The first stage in driver design is to check the brief or specification to ensure that it may be realised. How should we go about this? We find it easier to follow a practical example rather than a purely theoretical one, so to establish that the intended design is practical, let's use our proposed subwoofer.

The cabinet is 450 mm × 450 mm × 450 mm in 25 mm medium density fibreboard (MDF). As we saw earlier, that means an internal volume of 64 litres without taking account of bracing or the drive unit displacement.

Garner and Jackson's 'Theoretical and Practical Aspects of Loudspeaker Bass Unit Design' has a key graph showing the relationship between cabinet volume, –3 dB frequency, and sensitivity for a given bass alignment. Garner and Jackson's paper concentrates on just two alignments: the closed box and the B4 vented box, both critically damped.

We have produced a spreadsheet entitled 'Theoretical Bass Design' that allows us to check quickly whether or not a proposed specification is realisable.

Once a decision about the type of cabinet alignment has been reached, three main pieces of information are required to be entered into the spreadsheet.

We will continue with our closed box subwoofer design, setting it up as follows. First click the 'Closed' button, and then:

- Enter 86 (dB) as the minimum level required.
- Enter 86 (dB) as the maximum level required.
- Enter 32 (litres) as the lower volume available.
- Enter 100 (litres) as the maximum volume available.
- Optionally, we can set up the graph divisions as well these are the major and minor tick = 4 (Litre).
- Enter 20 (Hz) as the lower cut-off frequency.
- Enter 50 (Hz) as the upper cut-off frequency.
- Again, we can optionally set up the graph divisions as well: these are the major = 5 (Hz) and minor tick = 1 (Hz).

This allows us to see that with box volume of 64 litres, $R_1 = 1, Q_t = 0.7071$ and a sensitivity of 86 dB at 1 watt at 1 metre, the lowest –3 dB frequency for a closed box would be 38.6 Hz. This is shown as figure 16.1.

We can now calculate the rest of the drive unit parameters as follows:

$$R_1 = \frac{C_{ms}}{C_{mb}} \ or \ \frac{C_{as}}{C_{ab}} \ or \ \frac{V_{as}}{V_b} \tag{16.1}$$

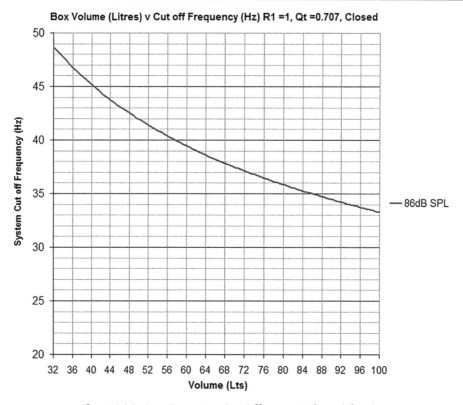

Figure 16.1: Low Frequency Cut-Off Versus Volume (Litres).

where

- C_{ms} = mechanical compliance of loudspeaker (m/N).
- C_{mb} = mechanical compliance of the air in the box (m/N).
- C_{as} = acoustical compliance of loudspeaker (m⁵/N).
- C_{ab} = acoustical compliance of the air in the box (m⁵/N).
- V_{as} = volume of acoustical compliance of loudspeaker (litres).
- V_b = volume of the air in the box (litres).

We have specified the box volume as 64 litres. As $R_1 = 1$, then (from equation (16.1) on page 71) the volume of acoustical compliance must also equal 64 litres.

The key to understanding this is that there are three equally valid ways of calculating R_1: mechanical, acoustical, or volume-based, and the actual values of each of these are related by the following equations:

$$C_{as} = \frac{V_{as}}{\rho_0 \cdot c^2 \cdot A^2} \tag{16.2}$$

$$C_{ab} = \frac{V_b}{\rho_0 \cdot c^2 \cdot A^2} \tag{16.3}$$

And naturally we can also calculate the resonance frequency:

$$F_s = \frac{1}{2 \cdot \pi \cdot \sqrt{M_{ms} \cdot C_{ms}}} \tag{16.4}$$

where V = volume of box in m³ (0.064 m³), ρ_0 = density of air (1.18 kg/m³), c = speed of sound in air (343 m/S), $\pi \approx 3.14159$, and A is the area of the driver.[1]

As the area depends upon the diameter, an equation for the area of a circle is also required. The following equations can then be used to calculate further parameters.

$$\text{Area} = \pi \times \text{radius}^2 \tag{16.5}$$

$$\text{Area} = \pi \times \left(\frac{\text{diameter}}{2} \right)^2 \tag{16.6}$$

$$C_{ms} = C_{ab} \cdot R1 \tag{16.7}$$

$$M_{ms} = \frac{1}{4\pi^2 \cdot Fs^2 \cdot C_{ms}} \tag{16.8}$$

$$B \cdot l = \sqrt{\frac{(2 \cdot \pi \cdot Fs \cdot M_{ms} \cdot R_e)}{Q_{es}}} \tag{16.9}$$

$$\frac{(B \cdot l)^2}{R_e} = \frac{2 \cdot \pi \cdot Fs \cdot M_{ms}}{Q_{es}} \tag{16.10}$$

$$Q_{es} = \frac{R_e}{(B \cdot l)^2} \cdot \sqrt{\frac{M_{ms}}{C_{ms}}} \tag{16.11}$$

$$Q_{ms} = \frac{1}{R_{ms}} \cdot \sqrt{\frac{M_{ms}}{C_{ms}}} \tag{16.12}$$

$$\frac{1}{Q_{ts}} = \frac{1}{Q_{es}} + \frac{1}{Q_{ms}} \tag{16.13}$$

$$Q_{ts} = \frac{Q_{ms} \cdot Q_{es}}{Q_{ms} + Q_{es}} \tag{16.14}$$

As a check we can calculate

$$Q_{es} = \frac{2 \cdot \pi \cdot Fs \cdot R_e \cdot M_{ms}}{(B \cdot l)^2} \tag{16.15}$$

The sensitivity E can be calculated from

$$E = 10 \cdot \log 10 \cdot \frac{p0}{2 \cdot \pi \cdot c} \cdot \frac{(B \cdot l^2)}{R_e} \cdot \frac{S_d^2}{M_{ms}^2} \tag{16.16}$$

To turn this into a dB SPL² requires a constant that depends on the loading into which the loudspeaker is radiating. For an infinite baffle, this is equal to 112.1 dB.

$$SPL \, (dB) = 112.1 + 10 \cdot \log 10 \cdot \frac{p_0}{2 \cdot \pi \cdot c} \cdot \frac{(B \cdot l^2)}{R_e} \cdot \frac{S_d^2}{M_{ms}^2} \tag{16.17}$$

From these equations (having first settled on a box alignment such as a closed box), we know $R_1 = 1$ & $Q_t = 0.707$, $V_b = 64$ litres or 0.064m^3, diameter $= 348\,\text{mm}$.

So $A = 0.095\text{m}^2$ and $C_{mb} = 0.000050958\,\text{m/N}$, so $C_{ms} = 0.000050958\,\text{m/N}$ or $0.050958\,\text{mm/N}$, which is a very small number.[3] So, it is very stiff. However, as $F_s = 38\,\text{Hz}$ and the moving mass $(M_{ms}) = 0.2944\,\text{kg}$ or nearly $1/3\,\text{kg}$ is quite heavy, it needs to be so.[4]

The overall power of the motor unit or 'shove' is $(B \cdot l)^2 / R_e = 170.29$. In practice, this is quite reasonable for a modern design of subwoofer driver with a typical $3.5\,\Omega$ D.C. resistance as this gives us a Bl value of $24\,\text{Tm}$, which is quite achievable.

Let us go back to our original specification. Right away there is a problem, as we did not actually have a meaningful specification. We said how loud we wanted it to be with 1 W input, but we didn't specify the output that we wanted from the system.

Subwoofers today are nearly always designed as systems, with both the loudspeaker and its amplifier considered together with an electronic crossover/equalisation.

So let us update our specification:

- Closed box cabinet 450 mm × 450 mm × 450 mm of 25 mm MDF.
- Driver 348 mm effective diameter.
- To use a 500 W rms amplifier (rated to deliver this output continuously).
- System to produce 110 dB SPL @ 20 Hz, harmonic distortion less than 10%.
- Electronic crossover/equalisation.
- System $Q = 0.707$.

Our problem is that this driver is only 86 dB at 38 Hz (1 W/1 m). But according to the specification, we need 110 dB at 20 Hz @ 1 m. So therefore, we need to find 34 dB (110 dB – 86 dB).

Since the amplifier is 500a,W it can produce 45 V rms versus a 1 W (or 2.83 V rms), which is a difference of about 24 dB.

So, we are around 10 dB down on maximum level. However, in practice a subwoofer is normally used near the floor (+6 dB) and often near a wall too (+6 dB). So in practice, we should just be able to meet the specification, although we shall see that this is not the case when we get to large signals design later.

As we have an active system, however, we can use a *Linkwitz transform* (named after Siegfried Linkwitz [2], who invented the concept). This allows a compensating active crossover that feeds the power amplifier and straightens out the frequency response, boosting and cutting the signal as required to meet a target response by producing an inverse curve, which 'corrects' for the changed system Q and a response deviation from the desired curve. We discuss this further in chapter 28, Linkwitz Transform, on page 127.

For our subwoofer, we now know that we can theoretically reach the required sound pressure level. ('Theoretically' because many other things can and do affect the sound at high power levels, as we saw in the chapter on large signal parameters.)

However, as a benefit of having investigated these parameters, we now have some key information that we can use in the rest of the design. At this point, we are ready to start a driver simulation. (In fact we already have the magnet and voice coil.) We like to begin by using a 2D axisymmetric model, for many reasons including:

- You can clearly see any clearance issues.
- You can see the position or offsets of the motor unit, cone/surround, or dust cap.
- You can take the final geometry to run an axisymmetric FEA model.
- It is a very efficient modelling method, using few degrees of freedom to describe the model accurately.

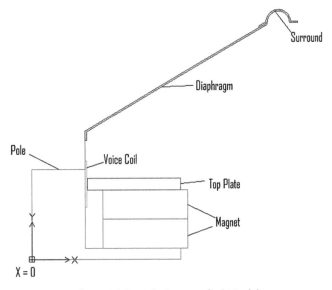

Figure 16.2: Axis Symmetrical Model.

So what is an axisymmetric model?

As you can see in figure 16.2, it is a fairly simple drawing. By convention it is normally drawn from $X = 0$ to the right, or positive X values. Initially it just looks like a cutaway section through the central axis, and it is. However, we completely discard any negative X values, so this now becomes a specialised type of 3D model: We can rotate this around the $X = 0$ axis, and this becomes a solid model, as we will see later.

A question that needs to be asked at this point is: 'Is a 2D axisymmetric model sufficient?' The answer depends on what you need from the model. Let us remember the saying: 'All models are wrong but some are nevertheless useful'.

If we are after the overall response together with some estimation of the off-axis performance or the underlying electromagnetics then the answer is probably 'yes'. However, if we want to model non-symmetrical cone modes or rocking modes, then the answer will be 'no' and a fuller model, potentially up to a full 3D one, may be required.

What do you need? If a 2D axisymmetric model will suffice then all is well; if you need a full 3D model then please do not assume that you can use a 2D one just because you can sweep or revolve the 2D axisymmetric model into a 3D shape. That's because such a model has none of the axial-rotational fine detailed variation that a full 3D model can contain. So, it cannot help you in the way that a full 3D model can contain and model rotational non-symmetries such as:

- Flux density variations.
- Positions of the windings.
- Lead-out wire position and tension.
- Voice coil former offsets.
- Cone/surround asymmetries.
- Cone/surround twisting moments.
- Spider twisting moments, and so on.

Let us take a simplified example of a very badly offset pole and top plate from our earlier axisymmetric model.

We have shown this as *really, really* exaggerated, as figure 16.3 clearly shows that the flux distribution would be massively focused towards the right hand side!

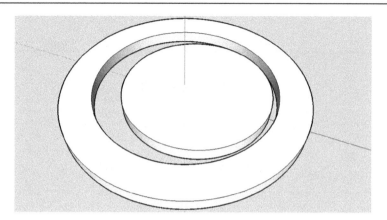

Figure 16.3: 3D Offset.

Could such a situation be effectively modelled with a 2D axisymmetric model? Of course not! If such a situation occurred it's very unlikely that the loudspeaker would still work. Such a situation would require a 3D model. Although a 2D planar model would capture some aspects of the magnetic field.

A 3D model could be used to help understand rocking modes, rub and buzz noises, or cone modes, and so forth. It really does depend on what one needs from a model, as well as the time and effort one is prepared to put into developing it. As another saying points out: 'There's no such thing as a free lunch'.

Notice that up until now we have not actually designed anything, but rather have started to build a structure of what our particular design is, what is in it, and what resources are required to complete it. This may include the complexity of the CAD and FEA models required for the next step(s).

Notes

1. We usually calculate this by including half of the surround diameter.
2. This is normally taken relative to 20 μPa, which is taken as 0 dB SPL.
3. In exponent format, 0.000050958 m/N is 50.958e–6 m/N.
4. Which is why subwoofer drivers never seem to move when you push them.

References

[1]   A. V. Garner and P. M. Jackson. "Theoretical and Practical Aspects of Loudspeaker Bass Unit Design". In: *Audio Engineering Society Convention 50*. Mar. 1975. www.aes.org/e-lib/browse.cfm?elib=2485.

[2]   *Linkwitz Transform*. www.linkwitzlab.com/filters.htm#9 (visited on 01/02/2018).

CHAPTER 17

Magnet

In this chapter we ask what a magnet is and look at magnetic flux lines and how to concentrate these into useful fields. We then discuss various modelling techniques and summarise the properties of some common magnetic materials.

All magnets produce what we call lines of force. These lines of force always flow from one pole to the opposite paired pole of a permanent bar magnet, as in figure 17.1.

Notice that although these flux lines never cross or intersect with each other, they are very close to each other inside the magnet and space themselves outside, while theoretically taking an infinite space (albeit at rapidly reducing flux levels). We use the interaction of a steady, permanent magnet flux and the changing magnetic field due to the current flowing through the coil in order to power the loudspeaker.

How do we make the transition between a bar magnet with flux lines going theoretically to infinity and a practical magnet system for a loudspeaker drive unit?

Many types of structure have been used for magnets. Rings and discs are the most common, but rods and arcs have been and are used for some designs. In order to analyse a structure, we need a drawing, typically of the axisymmetric model that we covered earlier. But it may be a full 3D model or a planar model which is simply a model looking in a single direction (and so can be used for bar magnets, commonly used in ribbon loudspeakers). A planar magnet system is shown in figure 17.2.

For the moment we will stick to a simple axisymmetric model (one that is fully symmetrical around a central axis). Typically, we will only cover a small amount of volume for the modelling, as magnetic fields fall off very quickly with distance; this helps keep the model small.

With most finite element programs it is imperative to select appropriate boundaries beyond which it will not attempt to model, and appropriate mesh densities so that the model has sufficient fine detail or resolution. Also, when

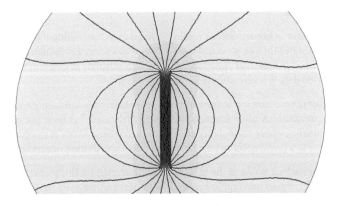

Figure 17.1: Bar Magnet Flux Lines.

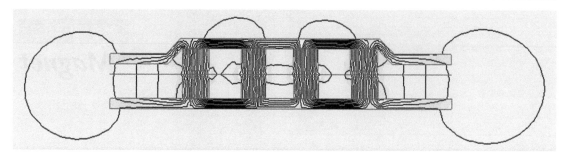

Figure 17.2: Planar Magnet Flux Lines.

producing CAD drawings, it is essential to ensure that lines that are supposed to join do so. Tiny gaps of 10^{-18} m or less can cause chaos in the model, either not solving or creating overlapping areas.

With some programs like FEMM [1], try to use small chamfers rather than radii in corners.[1] This can often reduce the mesh size dramatically. If you can select a larger mesh for a whole area without compromising results, say in a magnet, try it; always try to use symmetry about an axis to do so.

One trick that is often worthwhile is to examine a mesh visually to see if there are any areas or volumes where the mesh density is inappropriately high, as often this is an indication that the meshing routine is trying too hard to mesh finer and finer detail successively.

The better programs look after themselves to a degree, but all of these 'tricks' can drastically reduce the mesh size and give a much faster-running model. So if your model seems to be taking forever to produce a mesh, please check that you have not fallen foul of one or more of these problems. The program's mesh-maker might be trying to mesh impossibly fine detail, so if you can, check the number of nodes being meshed, which can often be seen quite clearly.

Sorting these problems out stops the meshing program working down to infinitely small levels and can speed up solutions dramatically.

Bearing these problems in mind, we will start by importing a clean, accurate and (simple to mesh) *.dxf drawing into FEMM or else drawing it up directly in FEMM. We will set up the initial conditions, boundaries and materials for the magnet structure, and ideally the voice-coil, together with a line through which you wish to model the magnetic gap B and $Bl(x)$ at the same time. We will then mesh the model and subsequently solve it, calling up a separate display or post processor to display the results.

We can also use the post-processor to export various results. Perhaps even controlling it by a scripting package like Lua [3] (Lua is integrated within FEMM); or we could use SciLab [5] or Octave [2] (which are free open source MATLAB [4] clones) if we wished to automate the process, or use it to calculate at different positions and frequencies. Or else we can use the exported data in a spreadsheet to calculate $Bl(x)$.

There are two types of magnet in common use at present time: ceramic ferrites and neodymium (otherwise known as rare earth types). However, previously, Alnico[2] magnets were extensively used, but their use is now largely restricted to high-temperature environments, where their ability to operate even when red hot is unequalled by any other magnetic material. This is due to a Curie point (where permanent demagnetisation occurs) of around 800 °C.

Ferrites are made from iron ore, which is one of the most common materials on this planet. (As a matter of fact, ferrite magnets are a by-product of steel making. In England, both steel and magnets were made in the North East.) Current ferrite magnets typically have a remanence of around 300 mT with a Curie point of about 600 °C.

Neodynium rare earth magnet materials are supposedly just that. However, that is not the reason for the currently high costs; that is primarily down to artificially limited supply. The world's supply now seems to come almost exclusively from China, and the Chinese Government impose export quotas to restrict supplies going outside the country. The U.S.A. used to supply, but manufacture effectively ceased when China started. Neodymium magnets typically have a remanence of around 1300 mT. However the Curie point is only around 150 °C, so using them even at relatively usual temperatures can be a problem.

As far as using the various types is concerned, they are similar, with a few provisos:

- Ferrites are much less powerful by unit volume than rare earth and Alnico types.
- Ferrites and Alnico will operate successfully at much higher temperatures.
- Ferrites are good electrical insulators.
- Ferrites cost less.
- Rare earths and Alnico are usually good electrical conductors.
- Rare earths are easily demagnetised at higher temperatures.
- Rare earths are much more powerful by unit volume.
- Rare earths and Alnico generally cost more.

Notes

1. When a radius intersects with a plane, it theoretically never meets it, so some meshing programs can sometimes keep on trying smaller and smaller mesh sizes to fill a non-existent gap.
2. Alnico is an alloy containing aluminium (Al), nickel (Ni), and cobalt (Co).

References

[1]   *Finite Element Method Magnetics.* www.femm.info/wiki/HomePage (visited on 31/01/2018).
[2]   *GNU Octave.* www.gnu.org/software/octave/ (visited on 05/02/2018).
[3]   *Lua programming language.* www.lua.org/ (visited on 05/02/2018).
[4]   *MATLAB – Mathworks.* www.mathworks.com/ (visited on 05/02/2018).
[5]   *SciLab.* www.scilab.org/ (visited on 05/02/2018).

CHAPTER 18

Voice Coil

In some ways the voice coil is the heart of the motor unit and hence the loudspeaker. Get it wrong, and the speaker will not function as desired. So we look not only at the materials used for voice coils, but also some alternative methods for designing them so that they meet our requirements and specifications.

Looking at the voice coil, we could use FEMM, though that cannot tell us anything about packing density, or the number of turns for a given volume. However, FEMM can calculate resistance and inductance. Alternatively, we could design it using other tools ourselves and see other details.

Voice coils usually use copper, copper-covered aluminium (CCAW), or aluminium wire. Silver has the highest conductance[1] among any metal at normal temperatures, but is not often used in loudspeakers, mainly for reasons of cost and weight. Gold is a good conductor (third behind silver and copper), but it's also relatively heavy. It doesn't tarnish, so makes reliable mechanical connections, but its high cost explains why it is often used as a very thin coating.

Although any conductive material could in theory be used for a voice coil, it is almost unknown to find magnetic materials used for this purpose, even though in theory these could help to maintain a high flux density by removing the non-magnetic gap (such as air) that inevitably reduces the gap flux. The trouble with using magnetic conductive materials is that they have a strong tendency to go off centre and attach to the pole or the top plate, effectively seizing up the whole motor unit.

There is, however, one class of materials that can do exactly this without such problems: Ferrofluid. Originally developed by NASA to provide magnetic seals in a vacuum, this liquid is used in nearly all modern tweeters to help dissipate heat, but it also conducts magnetic force rather better than air.

From the knowledge of a material's conductance, we can work out the resistance per length. Basic mathematics determines that the circumference of a circle is three and a bit times its diameter; this three and a bit has been given a special name π from a knowledge that π is roughly 3.14519.[2] We therefore know that the circumference of a circle is 3.14159 × diameter, so we can design a voice coil accurately.

Years ago, we designed a DOS-based program 'VCoils.exe' that works similarly. This is shown as figure 18.1.[3]

Whichever method we use, we will end up with a voice coil specified as follows:

1. Former inside diameter.
2. Former thickness.
3. Former material.
4. Coil inside diameter = former inside diameter + 2 * former thickness.
5. Former length.
6. Material of winding.
7. Conductance of material/(m) or resistance/(m).
8. Nominal winding diameter and max winding diameter.
9. Wind Length, number of layers.
10. Number of turns.

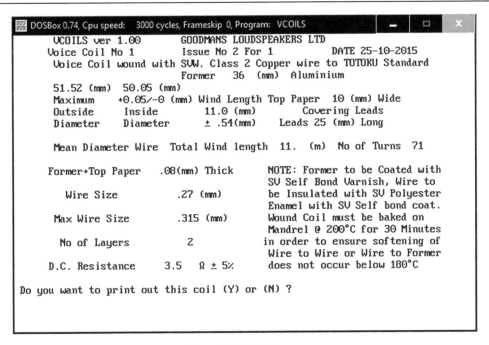

Figure 18.1: VCoils.exe.

11. Voice coil DCR.
12. Voice coil inductance.
13. Mass of the voice coil and former.

Quite a lot of factors need to be worked out. In fact, there is so much information we'd recommend a mini BOM or data table just for the voice coil and lead-out wires.

Notes

1. Conductance is the inverse of resistance.
2. π is pronounced as 'pie'.
3. Yes, it still ran, albeit under DOSBox 0.74. Please forgive the heading of Goodmans Loudspeakers Ltd, for whom it was originally written back in 1993.

CHAPTER 19

Force Factor Bl(x)

We saw previously that the voice coil is at the heart of a loudspeaker—however, Bl (B for flux density (T) and l for length of wire (m)) is how the voice coil translates the electrical current into the force that moves the loudspeaker. We also discuss in this chapter how Bl changes with displacement (x) and shove or $\frac{B \cdot l^2}{R_e}$.

From the magnetic flux calculated earlier and the voice coil design above, we can work out the Bl distribution over movement $Bl(x)$. And from using the tools in SpeakerPro or our spreadsheet, we have the underlying electromechanical parameters and already know the following ideal parameters:

1. Shove $= \frac{B \cdot l^2}{R_e}$.
2. Cone area $= S_d$.
3. Moving mass $= M_{ms}$.
4. Maximum SPL and -3 dB frequency.

From these we can estimate the cone movement or x_{max}, so given a magnetic gap size we can calculate the voice coil winding length and number of turns, inductance, and ultimately $\frac{(B \cdot l)^2}{R_e}$ and voice coil mass.

A designer should always try to close the loop by taking predicted outputs or simulations and compare these with the final results. That way, if there are errors or problems you can learn from them and prevent them reoccurring in the future.

However, we are getting slightly ahead of ourselves here as this comes under measurement, which we will cover in detail later.

This leads us to ask: 'How can we measure these quantities?' Bl is no problem as it can be calculated from the measured Thiele/Small parameters and a few equations set up in a spreadsheet.

Measuring $Bl(x)$ is trickier without a professional tool like the Klippel Analyser with the LSI module; however, the static force can be also be measured by using the whole voice coil to lift a known weight against gravity (9.81 Nm).

One could even build a force distance profile or $Bl(x)$ curve this way—we did it this way at Goodmans many years ago but it kept trying to burn out the voice coils so it was necessary to work quickly when making the measurements.

More importantly, what is $Bl(x)$? Well, just as $\frac{B \cdot l^2}{R_e}$ gives the shove or effective power of the motor unit, the (x) gives us the variation or change with respect to movement in x or normal direction of movement; $Bl(x)$ variations contribute directly to some distortion mechanisms.[1]

We do not propose to go into detail about these in this book but we can certainly try to minimise some problems by prediction of $Bl(x)$ for a proposed design and then by iteratively improving the design and running further simulations as required.

So, let us look at a $Bl(x)$ curve as measured by a Klippel Analyser. Figure 19.1 shows one of their example files.

Although difficult to measure, it is relatively easy to calculate, as shown in Chapter 21: Motor Unit on page 88 and in appendix D: FEMM Tutorial on page 207.

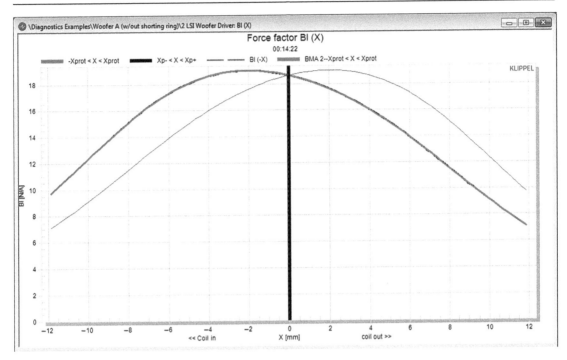

Figure 19.1: Klippel *Bl(x)* Curve.

Note

1. See the work of Wolfgang Klippel for further details.

CHAPTER 20

Inductance $L_e(x)$

Briefly, L_e is the inductive and the semi-inductive behaviour of a loudspeaker's impedance characteristic. We examine this first from a theoretical perspective, to understand the mechanisms behind the roll-off, then extend to the changes due to displacement of the $L_e(x)$ term.

Immediately we come back to our earlier small signal model of frequency response at high frequencies, where the response decreased in an odd fashion, dependent not just upon L_e but a weird combination of L_e and a semi-inductive component. So let us review L_e and how it affects a speaker's performance.

Up until now, the assumption has been that any high-frequency roll-off is due to an inductance or an inductance plus an additional inductor in parallel with a resistor. Examining the impedance rise of a real loudspeaker, the increase is typically not the 6 dB/octave that would be caused by an inductive component, but rather a much messier 3–3.5 dB/octave. The reason is down to an individual motor unit's physical characteristics, so what is happening? How can we model and measure it, and, more importantly, how can we use this knowledge?

Our key theories in this area come from John Vanderkooy's paper 'A Model of Loudspeaker Impedance incorporating Eddy Currents in the Pole' [8]. This explains these mechanisms by modelling an infinitely long pole and showing it under eddy current conditions. The skin depth is effectively constrained to the surface of the pole (decreasing with the square root of frequency), and thus the electrical conductivity of the pole and surrounding conductive materials is gradually reduced. This drastically changes the coupling away from a pure inductive model.

The problem with modelling this has long been one of capability, which until now has required high-end FEA modelling programs such as CEDRAT [2], COMSOL [3], or MagNet [1].

Recently, FEMM (which we will use later in our permanent magnetic analysis) was upgraded to allow us to model A.C. characteristics in the presence of a D.C. field. FEMM is available as a free download from [4].

The important thing is that we can now model the actual physical structure or a proposed structure and predict how we expect it to perform in reality. David Meeker, who wrote FEMM, has an extensive list of examples at: www.femm.info/wiki/Examples.

We are most interested in the four most recent ones:

- Blocked Impedance of a Loudspeaker (new).
- Blocked Impedance of a Loudspeaker (1 kHz) (new).
- Transient Loudspeaker Model (new).
- Transient Loudspeaker Model with Shorting Ring (new).

This describes how to use these new features to characterise the impedance of a loudspeaker across a range of positions and frequencies, with and without shorting rings.

To calculate or to measure $L_e(x)$ all we need to do is to calculate or measure L_e at different voice coil positions and then to plot the resulting curves. Fortunately, FEMM can also be driven automatically, so if we set up our model such that the voice coil is in a separate region, we can move just this part and calculate the results; likewise, we can run the analysis at different frequencies, exporting the results as required.

FEMM directly links to scripting language LUA [5] and can also be driven by GNU Octave [7], an open source equivalent of MATLAB [6]; either can plot $L_e(x)$.

We saw in Chapter 7: Small Signal Model on page 37 how L_e directly affects the high-frequency roll-off by interacting with R_e to provide an effective low pass filter. The (x) in $L_e(x)$ gives us the variation or change due to movement in the x direction.

Not surprisingly, $L_e(x)$ variations can contribute directly to some distortion mechanisms, though they tend to be less significant, or should we say less obvious, than those distortions due to $B_l(x)$ and $C_m s(x)$.

So let us look at a $L_e(x)$ curve as measured by a Klippel Analyser. Figure 20.1 shows one of their example files.

First we will update FEMM to the latest version[1] and skip ahead very slightly by opening a copy of our first loudspeaker motor unit. This will be called 'Subwoofer 1', as shown as figure 20.2.

We will go into the details of this model for the constant magnetic field and $Bl(x)$ in Chapter 21: Motor Unit on page 88. But here we need to focus on just the inductance calculation of the whole model, and it is better to do it with a realistic rather than a theoretical model.

We can now run a theoretical prediction of $L_e(x)$ by offsetting the coil at sufficient positions, as shown in figure 20.3, and we could also do so by running the simulation at different frequencies. Because we are doing this with the materials that will be used in a real motor unit, we will be able to cross-check these simulations with actual results later.

Running a series of simulations from –7 mm to +7 mm gives us the results shown in figure 20.4.[2] These results are clearly similar to the Klippel curves shown in figure 20.1.

From this curve we can see at least two things of significant interest to us. Firstly there is a very significant inductance; also, and critically, this inductance varies dramatically with the displacement, even over quite a small range of movement.

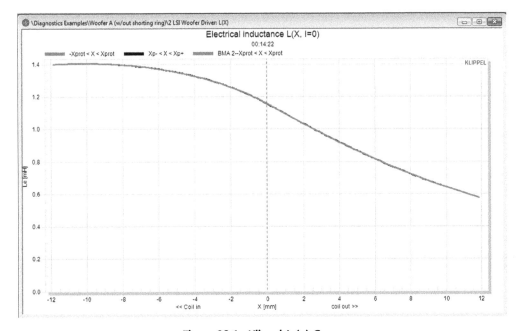

Figure 20.1: Klippel $L_e(x)$ Curve.

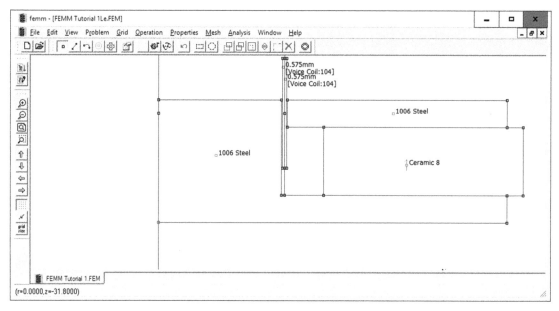

Figure 20.2: Subwoofer.

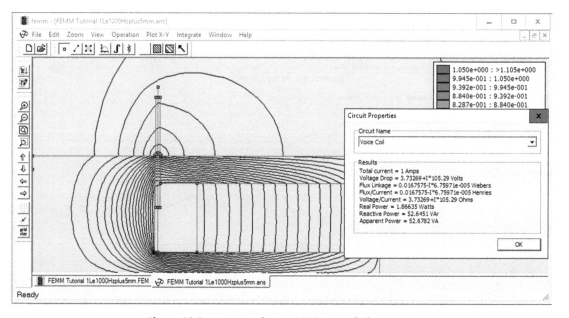

Figure 20.3: Motor Unit 1 at 1000 Hz and plus 5 mm.

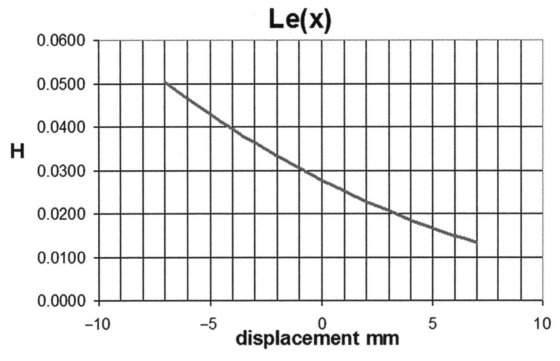

Figure 20.4: Motor Unit 1 $L_e(x)$ at 100 Hz.

Notes

1. Currently this is either FEMM 4.2 12 Jan 2016 (x64) or femm42bin_win32.
2. Details of this are shown in Appendix D: FEMM Tutorial on page 207.

References

[1] *2D/3D Electromagnetic Field Simulation.* www.infolytica.com/en/products/magnet/ (visited on 05/02/2018).

[2] *CEDRAT – FLUX.* https://altairhyperworks.com/product/flux (visited on 05/02/2018).

[3] *COMSOL Multiphysics.* www.comsol.com/ (visited on 31/01/2018).

[4] *Finite Element Method Magnetics.* www.femm.info/wiki/HomePage (visited on 31/01/2018).

[5] *Lua programming language.* www.lua.org/ (visited on 05/02/2018).

[6] *MATLAB – Mathworks.* www.mathworks.com/ (visited on 05/02/2018).

[7] *SciLab.* www.scilab.org/ (visited on 05/02/2018).

[8] John Vanderkooy. "A Model of Loudspeaker Driver Impedance Incorporating Eddy Currents in the Pole Structure". In: *J. Audio Eng. Soc* 37.3 (1989), pp. 119–128. www.aes.org/ e-lib/browse.cfm?elib=6100.

CHAPTER 21

Motor Unit

If as is arguably the case, the voice coil is the heart of a loudspeaker, then the motor unit is probably best described as the skeleton that holds everything together. Just as with the voice coil, get it wrong and no amount of tweaking with a crossover or DSP later will correct it, but get it right and things will go easier. So here we will develop a motor unit completely from start to finish.

One of the first things that needs to be decided with a motor unit are the sizes in particular the diameters and depths of various component parts like magnets, poles, and top plates.

We can design one from scratch using a CAD program like Draftsight or ProgeCAD to make a 2D axisymmetric *.dxf drawing that can be imported into FEMM for finite element analysis.

FEMM is an excellent program by David Meaker; it covers 2D axisymmetric and planar models. FEMM can work in the electromagnetic, electrostatic, or thermal domains and does an excellent job for electromagnetic models. We have cross-checked FEMM against actual designs for over 20 years and found it to be accurate to ±5%.

Although it's a bit crude by full draughting standards, FEMM does include a basic data entry facility, which you can use directly to enter the coordinates of a motor unit.

As we saw earlier, the electromechanical parameters describe the underlying parameters including Bl and R_e, but before we can design the motor unit, we need a little more information; we can gather most of this from our specifications, though.

The next pieces of information we need to know are:

- The −3 dB frequency.
- The maximum sound pressure level we require at this frequency.
- The effective cone area.
- The cabinet type and loading conditions including the amount of ripple in the response expected from the design.

From these four factors, we can estimate the amount of travel or X_{max} that the motor unit will need to provide to the cone. Once we have X_{max}, we can then start the process of selecting or designing a motor unit suitable to meet these requirements and also to meet the other design criteria for the loudspeaker.

Notice that almost immediately we are using the large signal domain by taking note of X_{max}; this will show up very shortly in our $Bl(x)$ curves. What other criteria are important at this point?

Well, it is possible that we are designing a full range drive unit so it might be essential to ensure that we do not have too much voice coil inductance rolling off the high frequencies. Or we may be designing a bass driver that needs to match a midrange driver, or a two-way system that needs to match with a tweeter. I'm sure you get the idea; there are lots of different possibilities.

It's at this point your product specification moves from really useful to essential. I really, really hope you have pinned it down, or you may be forced to do so later.

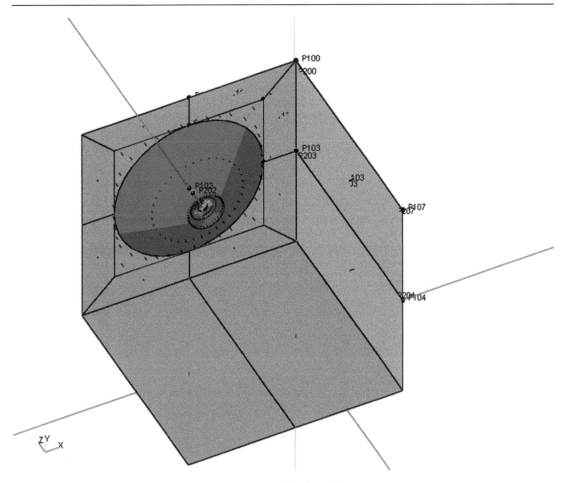

Figure 21.1: Driver in Cabinet.

But let us, for the moment, assume we are designing a subwoofer cabinet 450 mm × 450 mm × 450 mm and a driver to fit it. It is to be a closed box design; we have a 500 W amplifier available and will use a Linkwitz transform electronic filter/crossover to get the 'best' performance we can. The cabinet will be produced of 25 mm MDF, heavily braced and strengthened as shown in figure 21.1.

From our earlier calculations, we have the following parameters:

$F_s = 38$ Hz, $Bl = 25$ Tm, $R_e = 3.5$ ohms, $M_{ms} = 300$ gm, $S_d = 310$ cm^2, $X_{max} = 25$ mm, and -3 dB $= 27$ Hz.

As this is going to be a subwoofer driver, we are not too worried about any roll-off due to voice coil inductance. However, we need to have a power handling of at least 500 watts. We are going to need at least a 3-inch or 4-inch diameter voice coil; also, for an X_{max} of 25 mm, we are going to need a winding length of 25 mm plus the top plate thickness, so to reach 25 Tm we will probably need to use a 4-layer voice coil.

We can now search to see how close we can get to this with actual windings. If we use 0.575 mm copper wire in a 4-layer with a 32 mm wind length using a 75 mm ID giving a DCR of 3.46 ohms, this has 208 turns and the OD = 77.5 mm. Our spreadsheet figure 21.2 also says that for a 48 mm length of gap, this would equal 80 turns. So allowing 0.5 mm clearance inside and outside gives us a gap of around 78.0 mm − 74.5 mm, or 3.5 mm.

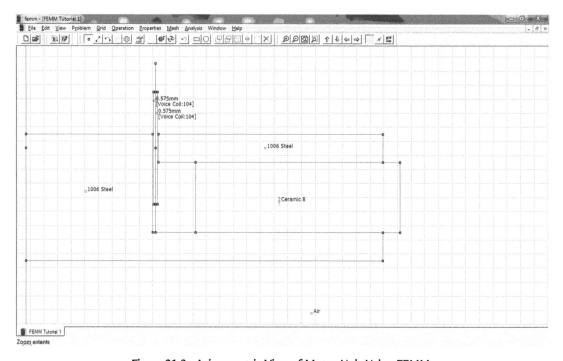

Dimensions					Wire Specifications							Calculations		
Inside Diameter (mm)	Inside Diameter (m)	Former Thickness (mm)	Wind Length (mm)	No of Layers (1-4)	Conductor Diameter (mm)	Conductor Width (mm)	Conductor Height (mm)	Conductivity (MS)	Packing Density (%)	Insulation Thickness (mm)	Insulation Thickness (m)	Wire spacing (m)	no. of whole turns per layer	Predicted DCR (Ω)
75	0.075	0.1	32	4	0.575			56	100%	0.0155	0.0000155	0.000606	52	3.46

Length of Gap Model (mm)	Wind Length (mm)	No of FEMM points	Total Turns	Coil OD
48	32	80	208	77.5

Figure 21.2: Theoretical *Bl* Voice Coil Design.

Figure 21.3: Axisymmetric View of Motor Unit Using FEMM.

We can now use FEMM to design the magnet structure. Let us start with a 220 mm OD × 100 mm ID × 20 mm thick ferrite magnet. Figure 21.3 shows an axisymmetric view of the structure.[1]

Notice how we have removed all the dimensions, leaving just the geometry of the parts from $X = 0$ and $Y = 0$, this model is axisymmetric about the Y axis. We have also located $X = 0$, as this also corresponds roughly the geometric centre of the magnetic gap and the middle of the voice coil.

At first sight, although an axisymmetric drawing looks like a 2D drawing it is *not* one; instead, it is a specialised type of 3D drawing. Therefore, it is essential that there are no negative X coordinates as the solid is assumed to revolve around the Y axis at $X = 0$. Strictly speaking, the Z axis should be exactly zero as well.

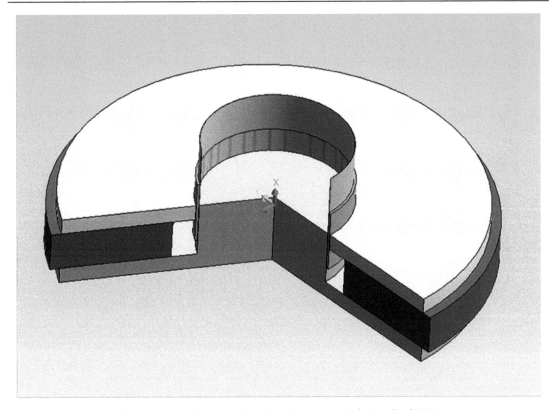

Figure 21.4: Axisymmetric Sub Driver Motor Unit Revolved 240°.

This is especially important if you are using parts drawn in a 3D program, as if they are even slightly off then various parts could become offset to each other, in the extreme case causing offsets as shown in the 3D Offset figure 16.3 on page 76.

To change it into a full 3D drawing all we need to do is to 'revolve' it around the Y axis; do this with a correctly generated axisymmetric drawing and it will look like a real loudspeaker. It's freaky the first time you do it correctly; see the results in figure 21.4.

In the 2D axisymmetric drawing, figure 21.3, we have included both the voice coil and drawn a line through the motor unit corresponding to the voice coil travel. By selecting this, exporting the flux along this line and by making some calculations in a spreadsheet, we can generate the $Bl(x)$ curve. Assuming an 8 mm thick top plate, we can estimate the flux required to meet our target of 25 Tm.

We will take you through just the main steps and analysis here. Please go to the FEMM and Theoretical Bl Tutorials in Appendix D on page 207 and Appendix K on page 319 for in-depth details of using these tools to run this type of design and analysis.

The first thing we will see after running an FEA analysis of the motor unit is a beautiful plot showing the magnetic flux distribution throughout the motor unit as shown in figure 21.5. With experience, this may give you enough information to go on; the rest of us will probably need to go a little further.

When we look at this, we can clearly see significant saturation in both the top plate and the back plate (this shows as the red and purple flux colours). So clearly some more work is required; however it's hard to see what else, if anything, needs to be done.

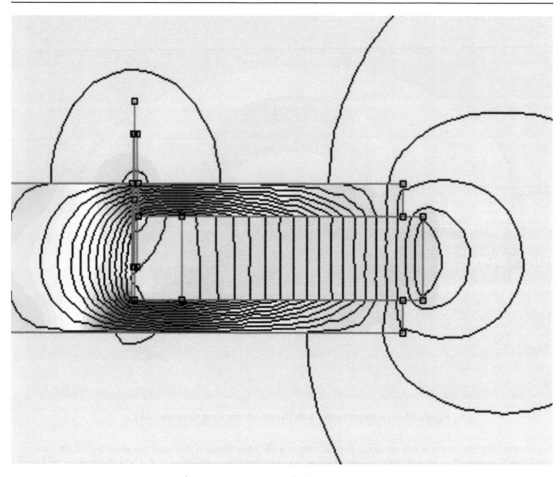

Figure 21.5: Motor Unit Simulated Field.

Our next stage is to analyse this data. We will take the flux line drawn through the voice coil/magnetic gap. This is shown in figure 21.6.

Further, we will take the flux line drawn through the voice coil/magnetic gap and plot it versus displacement of the voice coil, thus giving us the equivalent of the Klippel $Bl(x)$ curve. This is shown in figure 21.7.

We then calculate the difference between the simulated $Bl(x)$ curves giving us the asymmetry curves; these are shown in figure 21.8.

The important point is that we have produced these simulations without having to build anything physically—so we have a tremendous opportunity to try out new ideas and concepts without risk.

Now we are in a position to draw some conclusions:

- We have not achieved our 25 mm X_{max}.
- Our $Bl(x)$ is nearer to ±6 mm, reasonably flat, only approaching ±12 mm at 50% Bl; however, this is good, as otherwise magnet thickness would not let the voice coil move to the inward direction without hitting the back plate.

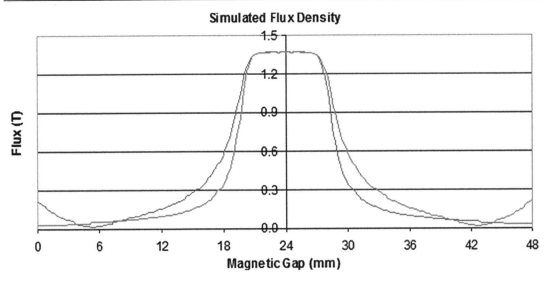

Figure 21.6: Motor Unit Simulated Flux.

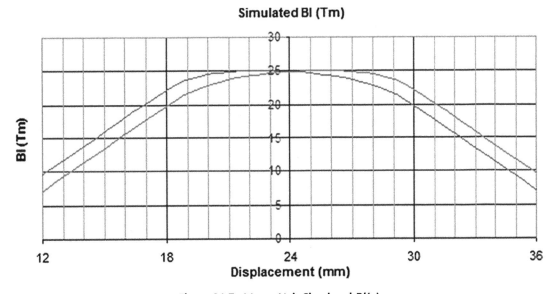

Figure 21.7: Motor Unit Simulated $Bl(x)$.

- There is significant asymmetry in the $Bl(x)$ curve.
- There is significant saturation showing in the back plate.

Okay, now that we know what is wrong or how we have differed from our initial targets and specifications, what can we do to correct these problems?

- We need a deeper magnet, so let's use two 20 mm thick magnets instead of one.
- The back plate is showing significant saturation, so let's increase its thickness from 8 mm to 15 mm.
- There is significant non-symmetry, so let's experiment with raising the pole height by 6 mm.

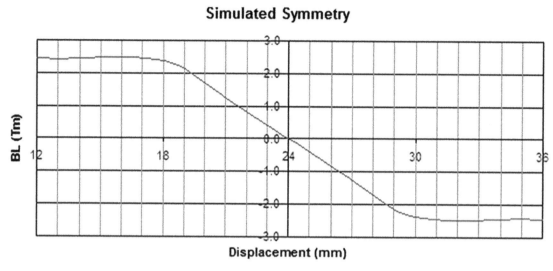

Figure 21.8: Motor Unit Simulated Asymmetry *Bl(x)*.

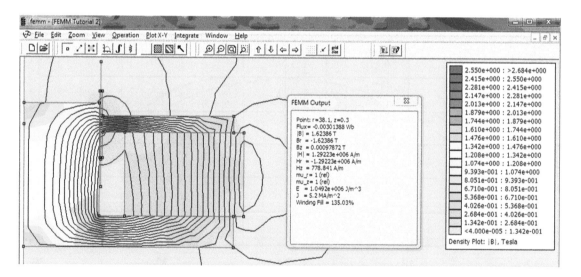

Figure 21.9: Motor Unit 2 Simulated Flux.

We can do this easily by making a copy of our FEMM model, highlighting the appropriate nodes, and moving them to the desired locations. The simulated flux is shown in figure 21.9.

Again we could revolve it to check if it looks correct—often a good idea as problems tend to show up more clearly when you look at them in a different way. The revolved model is shown as figure 21.10.

Again we will export the simulated flux and import into the spreadsheet. The results are shown in figure 21.11.

Whilst the simulated symmetry is shown on figure 21.12, we have a higher flux less saturation and better non-symmetry.

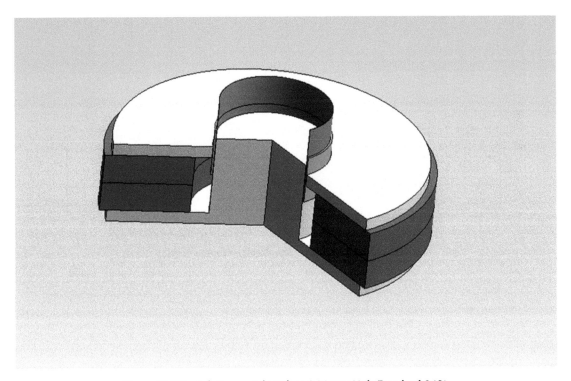

Figure 21.10: Axis Symmetric Driver 2 Motor Unit Revolved 240°.

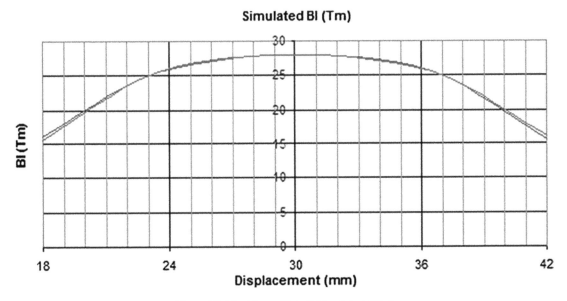

Figure 21.11: Motor Unit 2 Simulated *Bl*(*x*).

Simulated Symmetry

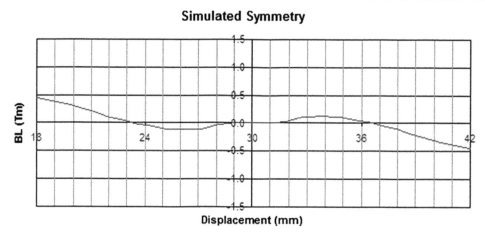

Figure 21.12: Motor Unit 2 Simulated Asymmetry *Bl(x).*

We now have a slightly higher *Bl*—which we should be able to cope with—less saturation, better symmetry, better *X_{max}* and more mechanical movement available. Job done for the moment.

Note

1. Conventionally this appears as a right half section; it is rotationally symmetric around the vertical axis.

PART VI

FEA, BEM, and Integration

CHAPTER 22

Material Specifications

In this chapter we discuss various types of material specifications: what they describe and why they matter. We then reference sources of further information.

What are material specifications? These are standardised descriptions of various physical material characteristics; examples could include:

- Electrical:
    - Resistance (measured in $\Omega\ m^3$).
    - Conductivity (measured in Siemens and the inverse of resistance, above).
- Magnetic:
    - Permeability the ability of a material to support a magnetic field within itself (measured in Henries per meter) $(H\ .\ m^{-1})$.
    - Remanence is the flux density B_r (Once a magnet is saturated it is measured in Tesla's) Wb/m^2 (T).
    - Coercivity (the ability to withstand external magnetic field without becoming demagnetised) (measured in oersteds or ampere/metre units) and denoted Hc.
- Physical:
    - Mass (measured in kg).
    - Young's modulus (tensile, compressive, and yield are all measures of stress to stain); Chapter 6 on page 28 is devoted to it.
    - Poisson's ratio (the negative ratio of traverse to axial strain). If a material is compressed in one direction it tends to expand in the other two directions. For most materials this is a constant between 0.3 and 0.5, though some special cases are different.
    - Damping (a dimensionless measure describing how changes decay after a disturbance).

Please note many of these material specifications are changed by variables such as:

- Temperature.
- Current.
- Frequency.
- Time.

Why do material specifications matter? Any or all of these parameters will change one or more aspects of a material's performance depending upon what it is being asked to do. For examples, a magnet with poor remanence and low saturation will probably not be very powerful, whereas a voice coil with very high resistivity will probably not pass much current.

Which material specifications matter most? All are important to one degree or another, the design trick being to combine appropriate materials with compatible material specifications to allow the desired function to take place.

How can we measure material specifications? We saw earlier how Young's modulus and damping can be measured; resistivity or conductance is generally well known and easily specified. Density (or weight) is rarely a problem today and neither is temperature, frequency, or time.

Take as an example the steels that are used in loudspeaker motor units. These are typically made from low carbon steels, and depending upon where you are in the world you may be able to get these as 1004, 1008, 1010 or various ASTM [1] types.

Although the internet has really been useful with the availability of mechanical data through sites like MatWeb [3] (in MatWeb search for 'Overview of materials for AISI 1000 Series'), many of important magnetic properties are often not readily available. Two major categories are listed here:

*   Soft magnetic materials or steels and magnetic conductors.
*   Hard magnetic materials or magnets.

Magweb **http://magweb.us/free-bh-curves/** has a wide range of curves of many steels available for download, while FEMM has many basic steel data built in as standard. BH data on magnets are also available online and a search for 'magnetic BH data' can reveal quite a lot of information but often in the form of multiple overlapped graphs. Many years ago, we used to use the old Mullard, an old British company name, later called the Philips Magnetic Databook. A modern example would be Intemag [2].

Both hard and soft magnetic data are really critical areas both for the actual loudspeaker and also for any modelling.

References

[1]   *ASTM Standards.* www.astm.org/Standards/steel-standards.html (visited on 05/02/2018).
[2]   *Intermag Magnetic Properties.* www.intemag.com/magnetic_properties.html (visited on 06/02/2018).
[3]   *MatWeb—the Online Materials Information Resource.* www.matweb.com/search/DataSheet.aspx (visited on 06/02/2018).

CHAPTER 23

Mechanical Finite Element Analysis

In this chapter we briefly describe how FEA works in non-mathematical language. We then discuss what is significant in a model and what is not, what is the level of appropriate complexity which we need to model, and only then do we start to build a model taking as inputs various parameters we worked out earlier.

We will use Mecway [2] for the mechanical modelling, calculating the displacements, eigenvectors, and modal shape(s) of the chosen cone and surround.

If we export the results in ASCII format, we could in theory use ABEC [1] and VACS [3] to display the predicted near-field and far-field frequency response by using our knowledge of the Thiele/Small parameters and sensitivity equations. Later, if so desired, we could get our final driver scanned by a laser to close the loop.

We can also use the CAD and FEA tools to predict the modes and overall compliance of the spider (used to ensure centralisation) and the dust cap (if used).

Earlier, in the chapter on motor units, we covered the use of FEA in magnetic analysis using FEMM, this is a specialised tool that uses FEA to solve particular problems.

However, FEA is itself a general technique. Let's assume we have drawn:

- A line, or;
- An enclosed shape, or;
- An enclosed volume.

Logically, any of these examples must be able to be described mathematically (have an equation); however, depending upon how elaborate the line, shape, or volume is, this equation might be *very* complicated.

FEA swaps this mathematical complexity for a much simpler model or equation, albeit one that must be performed many times (each time these calculations are made they are known as the degrees of freedom).

Without going into the mathematical details, FEA does this by taking the difference in a result calculated from one point or area or volume to another *very, very close by*, so as these results are taken from points or areas or volumes that are *very, very close* to each other these results should be very, very close to each other. Logically, if small enough differences between the points, areas, or volumes are used, these results will tend to become infinitesimal.

In the theoretical limit, the differences between successive results would be *zero*. Therefore, as successive results from a simulation are so similar to each other a very simple equation can describe them accurately.[1]

As we indicated above, there are specific or specialised programs and general ones. Another thing that tends to separate the two is the amount of knowledge and data that one has to enter into them.

Remember, FEA is a modelling technique, and as such it is not and cannot be *totally* accurate. There is a saying: 'All models are wrong, but some are useful'. I firmly believe this should be your guide here. One thing here is not trying to model everything: rather, keeping to an appropriate level of detail.

I know that many people will argue here that the whole point of FEA and such tools is to increase modelling accuracy and will cite the fact that our scientific methodology allows us to precisely calculate orbital mechanics and the like, and to do so with very high precision—and this is true. However, we do not know everything even here, and in my view, it is essential to focus on the things that are most important, rather than attempting to model everything in the universe.

We discussed this earlier when we thought about making a model of the world. So what would be the most important thing to model?

- A microbe?
- An individual person?
- A mountain?
- A continent?
- The whole globe?

Obviously, in this case, it's the whole globe that would dominate a model of the world. Even a continent or the tallest mountain is insignificant when compared to the whole world.

This is an essential concept to remember when applied to any model, but it is a *vital* concept when it comes to finite element or boundary element modelling.

You could take your full CAD model(s) and build your FEA model on these, and many programs include import capabilities from various CAD and 3D CAD formats; although it is tempting to use these, we would strongly advise against this, our reasoning being that the full 2D or 3D CAD model and drawings will inevitably have a lot of detail, sometimes fixings, dimensions, perhaps tables. This complexity can mean that lines do not meet accurately—almost inevitably, FEA programs need clean and clear data fed into them. So if you want to import from your CAD drawings, a lot of clean-up is often necessary, or modifications are required to make the model work effectively or even at all.

Instead, we would strongly recommend either constructing node by node, or using a simple and clean *.dxf file to build the FEA model, perhaps even generated from an FEA package like FEMM, which can then be exported into your main CAD drawing(s). After you have your final results, actually performing as you would wish, these then become the underlying drawing template(s). We want or need to model the acoustic output of a loudspeaker design, or—to be more precise—we wish to model the acoustic output of a proposed loudspeaker design before we have committed to producing it.

What can we do already, and therefore what remains to be done?

- We can model the overall low frequency response and sensitivity: Thiele/Small.
- We can model the motor unit (voice coil and magnet): FEMM.
- We need to design our proposed parts:
  - Voice coil.
  - Former.
  - Cone.
  - Dust cap.
  - Surround.
  - Spider or damper/centring.
- We need then to model the displacement/velocity distribution.
- We then need to model the acoustic response.

So for the moment we shall concentrate on these first components, which are present in practically all current loudspeakers in one form or another.[2]

Let us begin as follows:

- Draw up axisymmetric versions of these parts.
- Decide on the materials we will use and find their relevant properties.

- Draw up a combined minimal assembly—*not* the full loudspeaker.
- Draw these up in our chosen FEA program or MecWay, or produce a 3D model.
- Create an axisymmetric or a 3D model as appropriate.

Only now are we ready to really think about the finite element model itself:

- Decide on the appropriate analysis type(s).
- We then need to verify if the nodes and corresponding elements are correct.
- We then need to use the correct element types depending upon the model.
- Assign the correct material properties to the various elements.
- Assign the required loading and constraints to the model.
- Solve the model.
- Run the post-processing—look at the results to check if they are reasonable.
- Export the results and cross correlate with lumped parameter models.
- Export mechanical vibrational results in Klippel format.
- Import mechanical vibrational results in Klippel format into ABEC and display or plot the results.

Then, finally, we can compare overall results with our desired or design goals.

23.1 Begin Modelling

So for the mechanical modelling, we need to start by designing the following parts:

- Voice coil.
- Former.
- Cone.
- Surround.
- Dust cap if we are using one.

Specifically at this point, we need to generate the coordinates required to generate an axisymmetric view of these parts.

Let's look at an axisymmetric drawing of our proposed subwoofer drive unit, figure 23.1.

Notice how we have removed all the dimensions, leaving just the geometry of the parts from $X = 0$ and $Y = 0$, this model is axisymmetrical about the Y axis. We have also located $X = 0$, as this also corresponds roughly the geometric centre of the magnetic gap and the middle of the voice coil.

As we saw in Chapter 21, an axisymmetric drawing is a specialised type of 3D drawing, so it is essential that there are no negative X coordinates, and, strictly speaking, the Z axis should be exactly zero.

To change it into a full 3D drawing we need to 'revolve' it around the Y axis. Do it correctly and we should see the results shown in figure 23.2 on page 104.

This looks great and with some tools we can conduct our analysis from these drawings. Sorry, we quite can't handle this level yet; we will simplify our model a bit further. So let us split this combined drawing into separate parts. Firstly we will remove all the motor unit parts, as we have completed our initial work on these, so we can concentrate on the moving parts.

From our earlier work on the motor unit, we know the voice coil. If we use 0.575 mm copper wire in a 4-layer with a 32 mm wind length using a 75 mm ID gives a DCR of 3.46 ohms; this has 208 turns and the OD = 77.5 mm.

The voice coil will use 37.5 mm and 38.75 mm radius, top at 16 mm and bottom at −16 mm. The coordinates will be (37.5, −16), (37.5, 16), (38.75, −16) and (38.75,16).

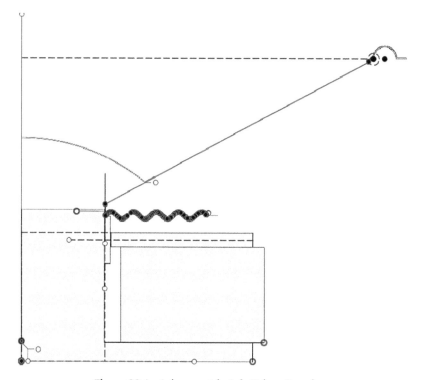

Figure 23.1: Axisymmetric Sub Driver Drawing.

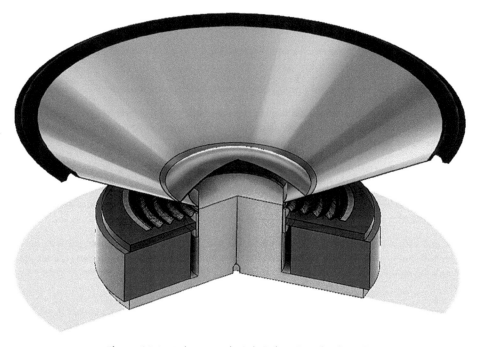

Figure 23.2: Axisymmetric Sub Driver Revolved 240°.

So for the voice coil:

X_{ivc} = internal radii winding = 37.5 mm or 0.0375 m.
X_{evc} = external radii winding = 38.75 mm or 0.03875 m.
Y_{tvc} = top of winding = 16 mm or 0.016 m.
Y_{bvc} = bottom of winding = −16 mm or −0.016 m.

For the former:
This is 36 mm + 16 mm = 52 mm or 0.052 m high.
It is currently 0.1 mm or 0.0001 m thick = tf.

Therefore, our coordinates need to be:

$X_{if} = X_{ivc} − t$
$X_{ef} = X_{ivc}$
$Y_{tf} = 0.036$
$Y_{bf} = −0.016$

The cone starts from the 6 mm below the top of the former; it is made from 1 mm thick material. It goes up by 109.61 − 36 = 73.61 mm or 0.07361 m to a radius of 166 mm or 0.166 m. So it goes out in the X direction by 166 − $Xivc$ or 166 − 37.5 = 127.5 mm or 0.1275 m, and it goes up in the Y direction by 0.07361 m, as we have already said.

Next, we have a half circle representing the surround; this has an 8 mm radius and it starts a little way down from the outer edge of the cone, but for the first model, we will assume that it starts directly and blends into the cone edge.

We will construct it by using a series of nodes with 15° between them, so we will need 180/15 or 12 nodes to describe these points. As we know the start and stop positions, we will only need to calculate the remaining ten nodes.

Procedure to start will be at the edge of the cone; the radius is 8 mm or 0.008 m so the centre point will be at $X_c = 166 + 8$, $Y_c = 109.61$.

Although this sounds complex when put into words, when we put it into a spreadsheet we can easily calculate these parameters using simple trigonometry. So for example, we know that the surround will in an axisymmetric model start at the far left, and for our purposes we will put it at exactly 180°. So taking the sine we can calculate the horizontal position and the cosine, the vertical position. See Appendix H on page 283.

Please notice that we are deliberately keeping the model as simple as possible. As Albert Einstein put it: 'Explanations should be as simple as necessary but not more so.'

There are many good reasons for this:

1. A simple model will run quicker than a more complex one.
2. It will be easier to see the important effects in a simple model.
3. It saves getting caught up in unimportant details that are not relevant.
4. It should not require detailed or (computationally) expensive 3D CAD Models.

We will use MecWay for the mechanical FEA. This can work in 2D, 2D axisymmetric, or 3D; we will use 2D axisymmetric dynamic to start with. This will enable us to apply a varying force to the voice coil in the Y direction whilst applying constraints to prevent movement in the X and Y directions at the outside edges of the surround, thus simulating a fixed immovable joint between the surround and the chassis.

The voice coil we will simulate as copper:

•   Density = 8700 kg/m³, Young's modulus = 117e9, Poisson's ratio = 0.33.

The cone, dust cap and former we will simulate as aluminium:

*   Density = 2700 kg/m³, Young's modulus = 69e9, Poisson's ratio = 0.33.

The surround we will simulate as rubber:

*   Density = 1100 kg/m³, Young's modulus = 0.001e9, Poisson's Ratio = 0.48.

The spider we will simulate as impregnated cloth:

*   Density = 1540kg/m³, Young's modulus = 0.43e9, Poisson's Ratio = 0.33.

MecWay's dynamic analysis uses steps in time, so we need to apply the suitable force in a suitable manner. For simplicity, we will use a sine wave and we will construct this in 15°steps from 0°to 345°; this gives us 24 steps and by setting the total time we can set the frequency that we want to analyse, as time equals 1/frequency. Therefore, we can run our analysis at any frequency we wish.

This sine wave is applied by a table, the first column being the time interval and the second column being the applied force.

One problem is drawing the very thin sections that often comprise loudspeaker components. One solution here is maybe to import from a dxf file. Natively, Mecway supports import from STEP and IGES, and also includes dxf import.

Notes

1. Some FEA programs show how close the results are from each other by calculating convergence.
2. Although a suspension is often missing in micro speakers.

References

[1]   *ABEC.* www.randteam.de/ABEC3/Index.html (visited on 02/02/2018).
[2]   *Mecway Finite Element Analysis for Windows.* http://mecway.com/ (visited on 31/01/2018).
[3]   *VACS.* www.randteam.de/VACS/Index.html (visited on 02/02/2018).

CHAPTER 24

Mechanical Compliance $C_{ms}(x)$

C_{ms} is the name we give to the mechanical compliance, and it is the ease with which a loudspeaker's moving parts are moved away from their nominal rest position. Compliance is measured in units of displacement versus force; in SI units, this is m/N. However, we are more interested in $K_{ms}(x)$, which is the inverse of $C_{ms}(x)$, as this describes how these parameters change with displacement or movement; it describes the *restoring force*, which keeps the loudspeaker's moving parts in the correct resting position.

In most loudspeakers there is nearly always something holding the moving parts in the correct geometric alignment.

That part or parts are often called suspensions, sometimes a roll surround and often one or more inner suspensions, somewhat confusingly called 'spiders': Would you like to know why?

Let us go back to figure 1.1 on page 4, showing details of the Rice and Kellogg loudspeaker. In this drawing there are two drawings, Fig. 2 and Fig. 3, showing how these early loudspeakers were held in the correct alignment—see it yet?

The key is the lines labelled 16; these are radially attached strings that are connected through a clamp coming down the pole and then through wires to the outside of the voice coil. Well, the clamp and wires looked sort of like a spider's web, and that is why the suspension or damper is or was often called the spider.

Strings, clamps, and wires have long since been replaced by more robust structures such as heat-formed tightly woven cloths; however, these still have their problems.

In our earlier small signal model we saw these modelled using capacitors, as these capacitors are effectively in parallel, so they are calculated as follows:

$$\text{Compliance Total} = \frac{C_1 \cdot C_2}{C_1 + C_2} \tag{24.1}$$

or alternately:

$$\frac{1}{C} = \frac{1}{C_1} + \frac{1}{C_2} + \frac{1}{C_n}. \tag{24.2}$$

We also saw in Chapter 7: Small Signal Model on page 37 that this capacitance acts with the inductance (actually the mass) to create a resonant circuit.

However, we can look at the mechanical compliance as a spring that resonates directly with the mechanical mass to give our system resonance; both views are equally valid.

At this point, we will look at $C_{ms}(x)$; just as C_{ms} directly affects the resonant frequency by interacting with the moving mass M_{ms}, the (x) gives us the variation or change versus movement in the x dimension or the direction of movement. $C_{ms}(x)$ variations can contribute directly to some distortion mechanisms. As we discussed earlier in many respects, we do not want a completely flat $C_{ms}(x)$ curve in a conventional loudspeaker system, as such a loudspeaker would be prone to self-destruction when driven outside its linear region. Therefore, a design with a completely flat $C_{ms}(x)$

would be prone to several major problems:

- There would be nothing to prevent the voice coil exceeding its normal excursion region (especially at resonance frequency where only the damping would exert any restraining force).
- The movement could easily produce D.C. offsets in the voice coil position (this is a cause of known and severe distortion).

One of the worst problems that can occur is when the compliance collapses at a particular displacement; the result of this is that the voice coil is suddenly free of its normal restraints and then moves to far inside or outside of its normal range. Sometimes this can cause 'popping noises': parts hitting other components and significant distortion or damage occurring.

So as a general rule, we wish to see some nonlinearity—if only to give some self-protection at high amplitudes or excursion.

How high excursion is defined depends entirely upon the use of the loudspeaker. For a subwoofer it may be ±25 mm, for a microspeaker ±2 mm, for an ultrasonic tweeter ±0.01 mm.

$C_{ms}(x)$ and its inverse $K_{ms}(x)$ are usually plotted as 2D graphs of $\pm$ displacement versus force. From the work of Wolfgang Klippel we know that by asymmetrical behaviour here can and does lead to various distortion mechanisms.

Also, the change in this restoring force can and does cause the resonant frequency to change at different displacements or excursions. This may or may not be a problem, but we should be aware of this. The question we need to ask in design is how can we prevent these problems occurring in our designs? Fortunately, our modern tools have advanced to the point where we can start to predict some of these problems.

Specifically, we can use FEA to model the force versus deflection curves for the surround and spider based on their geometry and material properties.

Let us look at a $C_{ms}(x)$ curve as measured by a Klippel analyser (figure 24.1)

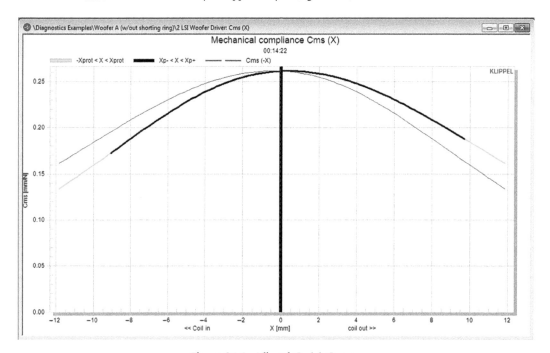

Figure 24.1: Klippel $C_{ms}(x)$ Curve.

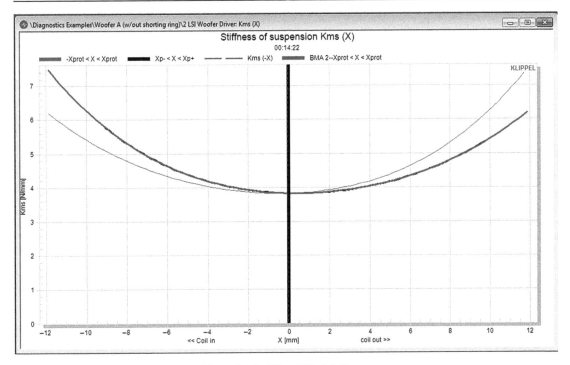

Figure 24.2: Klippel $K_{ms}(x)$ Curve.

and then at a $K_{ms}(x)$ curve as measured by a Klippel analyser (figure 24.2).

Ultimately, we could perhaps even predict the distortion contributions of such changes in the same way that the Klippel simulation module does, but this would be in the advanced class, and we will stop here, noting that the FEA should be able to help us at least visualise some of these details.

CHAPTER 25

Suspension

In this chapter we decide the essential requirements that a suspension must meet. We build an axisymmetric model of the proposed suspension and model one using a nonlinear static model, exporting the force–displacement results into a spreadsheet for visualisation.

Earlier we worked out the moving mass or the weight(s) of the various parts and distributed this appropriately. Now we need to work out the suspension for the loudspeaker. As we are designing a subwoofer this is a critical part of the design; let us consider what it needs to do in order to function well:

1.   It must allow free movement in the main direction of the applied force of ± 25 mm while preventing movement outside of this range.
2.   It must prevent any other side-to-side or rocking movements while doing so.
3.   While doing (1) and (2), it must hold the voice coil and cone surround assembly in the correct concentric alignment(s).
4.   While doing all of the above it needs to apply sufficient stiffness to the total moving mass such that the drive unit resonance is maintained at its correct value.
5.   The stiffness applied must remain well controlled over a range of ± 25 mm so that it does not cause significant nonlinearities and distortion.

25.1 Compliance $C_{ms}(x)$

Just as C_{ms} directly affects the resonant frequency by interacting with the moving mass M_{ms}, the (x) gives us the variation or change in movement in the x dimension in the direction of movement. $C_{ms}(x)$ variations can contribute directly to some distortion mechanisms.

As we discussed earlier, one of the worst problems that can occur is when the compliance collapses at a particular displacement. The result of this is that the voice coil is suddenly free of its normal restraints and then moves to far inside or outside of its normal range. Sometimes this can cause 'popping noises'; more typically it is the underlying cause of some types of significant distortion.

Wolfgang Klippel has made a detailed study of this and the Klippel analysers can directly measure both $C_{ms}(x)$ and consequent distortion. So now we can design effectively and minimise the 'black art' or 'suck it and see' approach of the past.

Here again, we will adopt the 'break it up into small pieces' approach. Let us start with the proposed geometry of a spider (suspension); we will import this into MecWay. We will then run a mechanical model but are really concerned here with its force versus deflection curve. (We saw earlier that this will identify some major contributing factors in this area.)

For our first step, we will create a potential design for a spider or suspension. We've taken the section through a one-inch spider and rescaled it to fit our subwoofer design. Then we cut off the left half as we will be creating an axisymmetric model (as shown in figure 25.1).

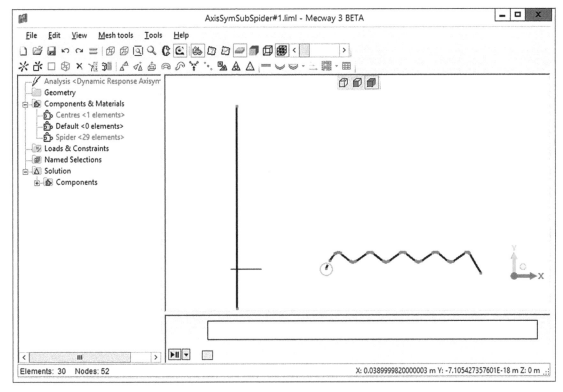

Figure 25.1: Axisymmetric Sub Spider.

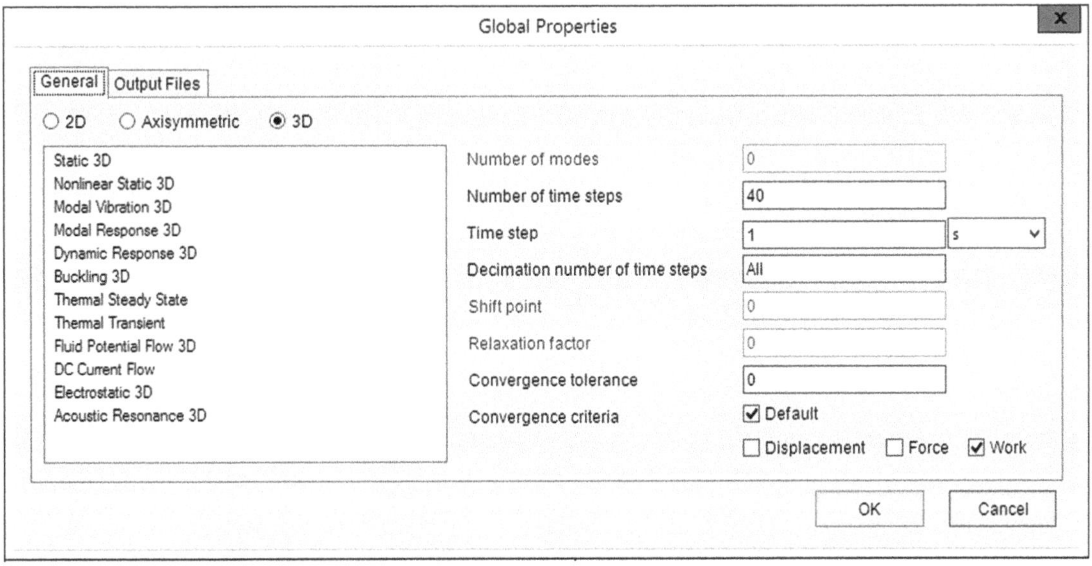

Figure 25.2: Global Properties.

We saw earlier that the $C_{ms}(x)$ is the compliance and $K_{ms}(x)$ is the stiffness. We will concentrate on the stiffness versus displacement curve. To simulate this, we will simply apply a known static force and calculate the resultant stiffness, then plot the curve between these values.

We have a confession to make. Even though we've been doing this for years, we *still* forgot that this is a nonlinear model, and we could not understand why all our results were completely *flat*.

The next step is to choose the correct modelling mode. We will use the nonlinear static 3D mode and need to right-click on 'Analysis<Dynamic Response Axisymmetric>', click 'Edit' and click the 3D button, selecting 'Nonlinear Static 3D'. We also need to enter the number and length of the time steps; we will use 40 and 1 s as shown in figure 25.2.

Before going any further, we need to convert the curved line, which represented our suspension in the axisymmetric view, into something suitable for a 3D model.

We will select all of the nodes and copy them 0.25 mm in the Z axis; this creates a very thin 3D shape rather than an infinitely thin section, thus cutting down the number of nodes required for the model (which can be extended later if necessary).

We then need to rebuild all the elements into ones suitable for use in a 3D model. We will choose hex8; originally they were quad4 elements. We will use Edit->Delete Elements, retaining nodes.

Our next step is then to create the new elements suited for 3D. Go to Mesh Tools->Create->Elements. The 'Add Element' dialogue will appear, as shown in figure 25.3.

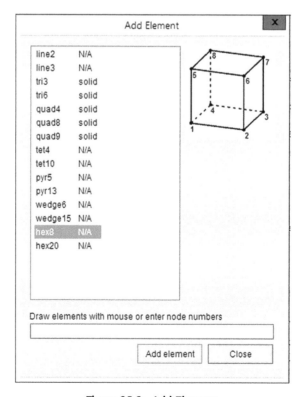

Figure 25.3: Add Element.

We will select and use hex8. We then zoom in and click on individual nodes one at a time, being careful to follow the order shown in figure 25.3. This process is shown for the first element in figure 25.4.

Notice that when each node in turn is selected, it is highlighted by a yellow circle.

The first element is shown by first selecting the elements and then an individual element and is shown in figure 25.5.

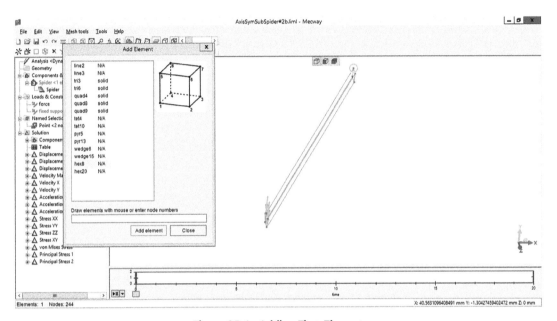

Figure 25.4: Adding First Element.

Figure 25.5: First Element.

We can then continue until all nodes are correctly connected with suitable elements. We will now switch back to our working model. An isometric view is shown in figure 25.6.

We can change the orientation at will by clicking on the X, Y, or Z axes or by holding the wheel button down on the mouse and moving the model around. Our next task will be to apply the forces and constraints to our model: Click on 'Loads & Constraints' (here it's already been done), then 'New Force'. We will apply this using a table in the Y direction, shown as figure 25.7.

We have applied a slowly ramping force from -0.01 N to $+0.01$ N; we can then use the FEA later to predict the displacement and hopefully plot the $K_{ms}(x)$ curve. We will now switch our attention to the outside of the suspension

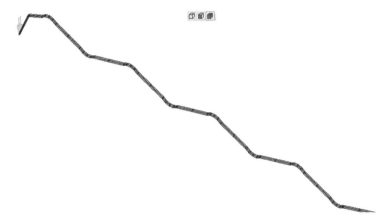

Figure 25.6: Suspension 3D Isometric View.

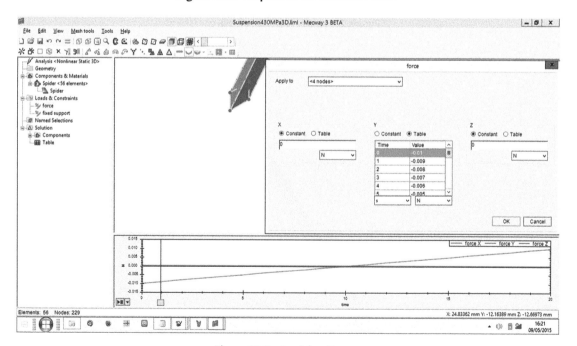

Figure 25.7: Applying Force.

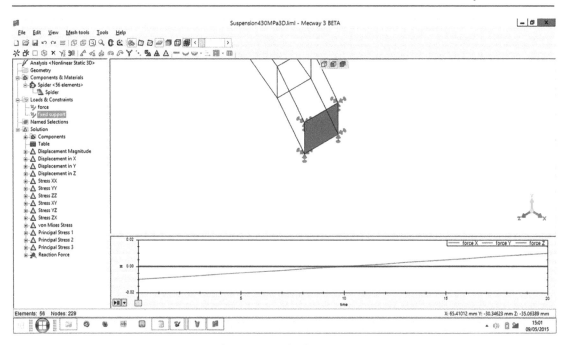

Figure 25.8: Fixed Support.

as we need to fix this and prevent it from moving. We will zoom in on the other end of the model as shown in figure 25.8.

As we can see, this support is fixed to a *face*. Provided that we have put all of the information in correctly, this model should now run. Press the 'Solve' button (which looks like a green '=' sign). If all is well, you should then see the 'Solver Convergence' dialogue, shown in figure 25.9.

This dialogue changes as the program goes through each step in turn and as it hunts for or converges on an accurate solution. Depending upon the size and complexity of the model and the number of steps required, this may take some considerable time.

This is why we take pains to reduce a model to its most basic form. Again, as Albert Einstein put it: 'Everything should be made as simple as possible but no simpler.' The caveat here is to ensure that we have not over-simplified the model but have only reduced it until it can be reduced no further without losing significant accuracy.

Why do I say this? Obviously an axisymmetric response model would have been simpler, but using it would have sacrificed nonlinear modelling which is essential in this case. When it goes green, it has finished and we can start to examine the results and see if they make sense. To keep it simple we will choose 'View->X-Y Orientation' and 'View->Fit to Window'; we will click on 'Force', and then on 'Displacement in Y'.

If we click on 'View undeformed', all looks good, but click on 'View deformed' and initially it looks crazy! We just need to reduce the amplification (in this case from 1 to 0.5). Then click on the undeformed shape as well as the deformed shape. When we click on the play button we can cycle through the simulation, as shown in figure 25.10.

Our next task is to extract this information by double clicking on the table. This is shown as figure 25.11.

If we enter a name for the CSV file and run it again, it automatically generates a CSV file. Such a CSV file can be plotted as the force versus deflection curve as shown in figure 25.12.

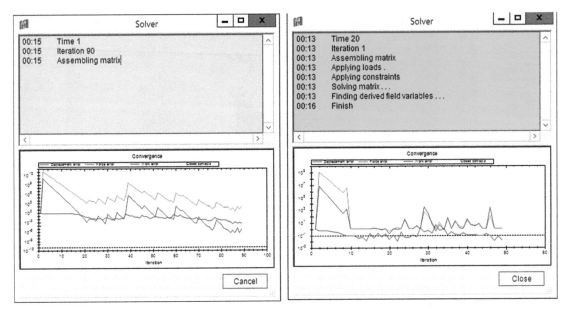

Figure 25.9: Solver Convergence.

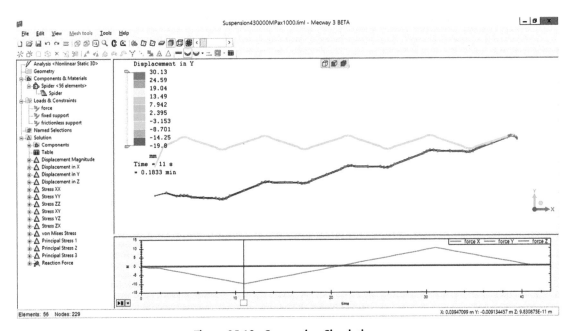

Figure 25.10: Suspension Simulation.

Figure 25.11: Suspension Table of Results.

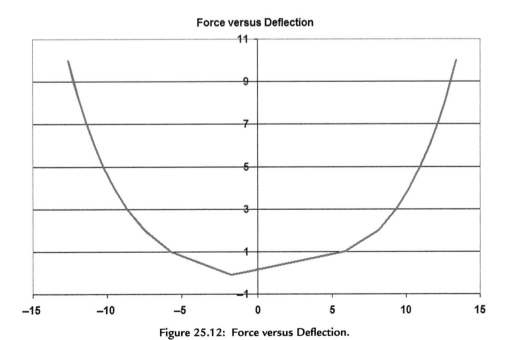

Figure 25.12: Force versus Deflection.

CHAPTER 26

Mechanical Simulations

This chapter discusses why we produce mechanical rather than acoustical simulations.

(i) The efficiency and coupling is usually so low that such simplification makes little difference. (ii) Programs like ABEC, which stands for Acoustic Boundary Element Calculator [1], can directly accept such mechanical data or calculations to produce the acoustic results we will need later anyway.

Earlier we used an FEA solver called FEMM for the motor unit; to solve this we needed to introduce an artificial boundary. We also saw how an FEA solver could be used to sort out the mechanical nonlinear forces required to understand the spider or suspension components.

We now need to work out which tools would be best to solve the mechanical to acoustical domain; the problem with using FEA is that FEA works most effectively in closed boundary situations—fine for the mechanical side but most acoustical problems are open or infinite. This leads to the use of boundary layers or the use of boundary elements or BEM solvers for the acoustical transmission between the loudspeaker and the ear.

For much of this book we will base this acoustic simulation work on ABEC from R&D-Team. This uses a BEM solver and is available in 32-bit and 64-bit versions for the Windows platform. (A detailed tutorial ABEC and VACS Tutorial is included as Appendix C on page 189.)

R&D-Team has collaborated with Klippel, and the Klippel scanning hardware and software has the capacity to import scanned data directly from a loudspeaker. We shall aim to format our mechanical simulation(s) in a similar fashion; this is called a *VibFile*. Thus we will be able to use ABEC later to simulate the acoustic output for us.

26.1 Cone and Surround

The next major task is designing the cone and surround assembly. We will split this into two parts. The first is a repeat of our earlier investigation on the suspension. Here we will just repeat the simulation but apply a known force to the inside of the surround and calculate the resulting displacement.

This will give us an equivalent $K_{ms}(x)$ curve, but for the surround alone. We then add this to the results earlier from the suspension to provide us with an estimate for the $K_{ms}(x)$ and its inverse $C_{ms}(x)$.

Next, we will turn our attention to the response of the cone or diaphragm; this is one of the most critical components when you get above the pistonic region. To be precise, this is the frequency region where the loudspeaker cone or diaphragm ceases to act as a unified whole, and where different areas of the structure start moving at different velocities and in various directions.

An often-asked question is surely that we want the whole cone to respond as one, don't we? The answer is both yes and no: Yes, because having the whole cone responding as one mass would simplify our job enormously in many respects, allowing a much simpler model. No, because if the whole cone responded as one part, then the frequency response would inevitably become mass controlled, resulting in a fall-off of the SPL with increasing frequency.

Another problem though with such a gradual roll-off in the radiating area (and thus a fairly flat acoustic output) is that almost inevitably the overall sensitivity and efficiency will be lower, as we have less area to produce our desired output (that is the SPL, of course).

Also, if the whole cone was acting as one it could (and probably would) start to 'beam' significantly. In other words, the sound would not spread out consistently with increasing frequency. This is called the directivity index (DI).

Earl Geddes [4] has made a thorough study of this subject, and the interested reader is directed to his work published on his website [2] and in his many papers published by the Audio Engineering Society.

So how could we go about predicting—or more accurately designing for—a particular response shape? Several alternatives are available ranging from the analytical methods of Frankort [3] to structural FEA. The latter path will be explored in this book.

We will build from scratch using structural FEA to generate the physical vibrations of a cone or surround assembly; we will format the output such that either ABEC or the Klippel scanning system [5] can recognise it and thus produce an amplitude versus frequency response of the driver and/or system.

As we (hopefully) will have generated cone surround vibration over the whole area, we should also be able to use this to produce either conventional polar responses or directivity response plots. These can be used to demonstrate the performance of the driver or system ideally before we build it. And it is at this stage when changes can be made with the least cost—often effectively free.

What do we need to model?

In the case of our subwoofer we are fortunate to have a simple and symmetrical structure, so we should not need to use a full 3D model of the driver. And as it is a subwoofer, it will be producing wavelengths much larger than itself as a consequence, so a simple 'pistonic' behaviour model should suffice. As the physical design is a cube, we can use symmetry in two planes to simplify both the cabinet design and the modelling.

So for our subwoofer model, we should be able to use MecWay and ABEC for the major part of the modelling. Our modelling will start again with the familiar 2D axisymmetric model, but this time it will include all of the major components:

1. Voice coil.
2. Voice coil former.
3. Suspension.
4. Cone or diaphragm.
5. Dust cap.
6. Surround.

Note: although the suspension is included here, it often has little direct effect on the sound produced by the loudspeaker driver. One modelling trick is to leave the suspension out of a mechanical–acoustical model; this may or may not make a significant difference.

As we would always suggest starting as simply as possible, we will leave the suspension out of our first model. We will now run a mechanical FEA model but can now look for additional information from its outputs:

- Displacement.
- Velocity.
- Acceleration.
- Modal behaviour.

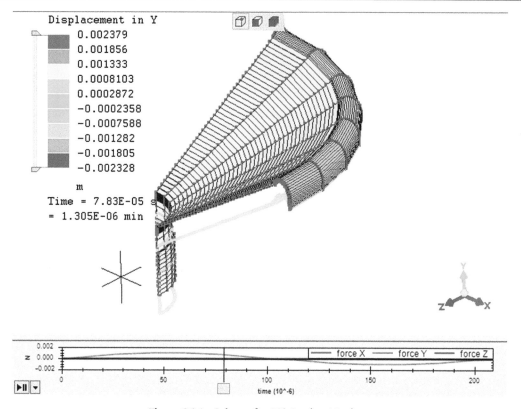

Figure 26.1: Subwoofer 90° Section Moving.

We can look at these at various outputs at different points in the model to understand how the underlying structure is moving when various forces are applied to the voice coil and how the structure responds to different frequencies as shown in figure 26.1.

A more detailed example is shown in Appendix H on page 283; a 90° section of this model is shown as figure H.24 on page 303.

References

[1]   *ABEC.* www.randteam.de/ABEC3/Index.html (visited on 02/02/2018).

[2]   *Earl Geddes LLC.* www.gedlee.com/ (visited on 06/02/2018).

[3]   F. J. M. Frankort. "Vibration Patterns and Radiation Behavior of Loudspeaker Cones". In: *J. Audio Eng. Soc* 26.9 (1978), pp. 609–622. www.aes.org/e-lib/browse.cfm?elib=3251.

[4]   Earl R. Geddes. "Acoustic waveguide theory". In: *J. Audio Eng. Soc* 37.7/8 (1989), pp. 554–569.

[5]   Wolfgang Klippel and Joachim Schlechter. "Measurement and Visualization of Loudspeaker Cone Vibration". In: *Audio Engineering Society Convention 121*. Oct. 2006. www.aes.org/e-lib/browse.cfm?elib=13716.

CHAPTER 27

PafLS

PafLS [1] takes a very different approach to loudspeaker simulation; rather than go into the detail of electromechanical simulation, it takes the key Thiele–Small or electromechanical parameters. These are than fed into a mechanical template that couples to the simulation engine behind PAFEC, avoiding the design complexity at the cost of limiting otherwise infinite choices.

PafLS was developed to provide a graphical user interface (GUI) for a PAFEC engine, specifically for the loudspeaker industry. A non-commercial version of this is available for £100, which fits it into the low-cost software category. It provides a fully coupled, script-driven mechanical-to-acoustical FEA modelling program. As discussed earlier, this means that the acoustical output is coupled to the mechanics (and vice versa).

Although the theoreticians may argue with us here in most cases, we feel that whether or not a model is fully coupled is not usually a severe problem.[1] This is because the efficiency and acoustic coupling in most loudspeakers is so poor that not having bidirectional coupling does not usually affect the simulation severely.

Rather than building a structure from the ground up, as with MecWay, PafLS takes a template-based approach to developing the underlying model. It can solve a limited range of standard loudspeaker shapes and forms and can vary a number of the inputs that define these by changing directly the underlying parameters, materials, shapes, and so forth, and mapping these into inputs that the underlying PAFEC FEA engine can directly understand.

PAFEC is a full commercial FEA software package. It's used quite widely in the loudspeaker industry for modelling drive units and so on, so is outside the scope of this book. Like many of the early general-purpose programs, it can be quite difficult to use. We first used PAFEC some 25 years ago at Goodmans, and must confess we found it difficult, as it required a detailed knowledge of the FEA process plus knowledge of which element types to use, where, and why.

In this respect, we remember it was very similar in operation to ANSYS and other early FEA programs. As we understand it, most of these developed out of the early FEA programs developed originally in the 1960s. In the 1960s and even on into the 1970s, computers were primarily used as number crunchers. Obviously they still do this, but back then visual displays were rarely used to demonstrate physical movements; simulation data was typically displayed in reams of printouts.

These were the products and programs of a different age, when compactness of code and efficiency of calculation was the only thing that mattered. Computers back then had tiny memories and storage capabilities, and were even weaker in terms of calculation abilities. By today's standards, one of the strangest things was the lack of any visual display, as television had been around for decades, but this was perhaps understandable as the calculations required to produce them were just not practical until relatively recently.

These FEA programs were originally programmed by the very people who would later run the simulations. So these people were well versed in the fundamental theory and had little if any need to ensure a clarity of approach, with the inputs typically being fed into the programs by scripts.

The underlying PAFEC engine then produces simulated responses, and by comparing these with the measured responses, one can refine a design in detail. The advantage is that some of the typical design choices, such as cone or diaphragm shapes, suspensions and surrounds, can be implemented.

The downside of the PafLS approach is of course that you are limited to designs that lie within the template structure, and these are limited to just a few of the infinite possibilities. But this does simplify the task dramatically.

What can it do well is axisymmetrical, mechanically-coupled acoustical simulations of a single driver. It does not do any magnetic modelling, nor can it handle multiple drivers, dual concentric, or other more unusual situations.

So it cannot model the voice coil or magnetic structure. Rather, these are treated as inputs in the same way that ABEC can also accept Thiele–Small or electromechanical parameters.

Neither can it simulate the external shape of, say, an enclosure (which ABEC can). We will see here how it handles our subwoofer design. (In Appendix J on page 310 we will go into more detail about how to use PafLS.)

We will start our design by adapting the LS-WE loudspeaker model to simulate our full subwoofer model, beginning with the structural layout tab. The various parameters are shown in graphical form as figure J.4 on page 312, the dimensions pointing to the individual parts that can be changed.

Selecting the 'Analysis' tab and 'Voltage Drive' is where we also input the blocked impedance or D.C.R, inductance L_e, and Bl. This is shown as figure 27.1.

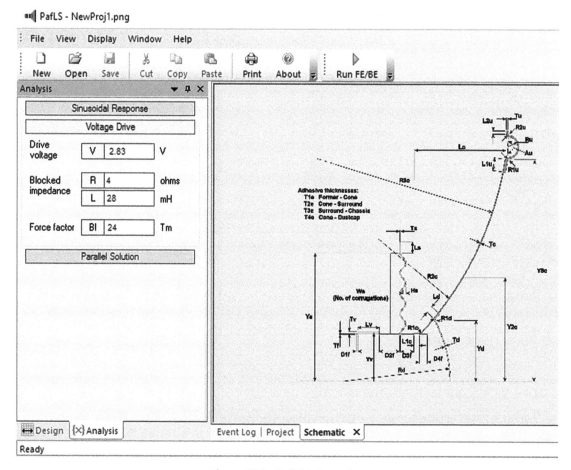

Figure 27.1: PafLS Schematic.

Other individual parameters that can actually be changed are shown for our subwoofer in the tables for the subcomponents in figure 27.2. We have entered the straight-sided cone as two large radii R2c and R3c of 0.99 metres each.

After the simulation has successfully run, you can then get access to the visual model that was represented in the tables. This is shown as figure 27.3.

The predicted SPL is shown as figure 27.4.[2]

The acoustic field pressures are shown in figure 27.5.

The electrical impedance is shown in figure 27.6.

The mesh used is shown as figure 27.7.

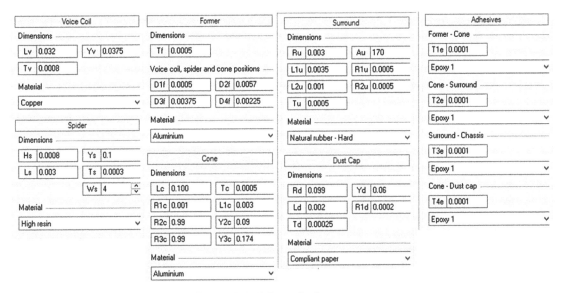

Figure 27.2: Subwoofer Parameters.

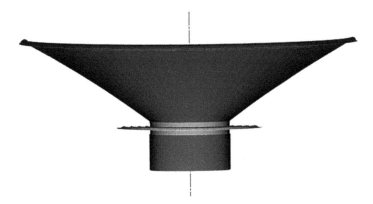

Figure 27.3: Structure of the Subwoofer Driver.

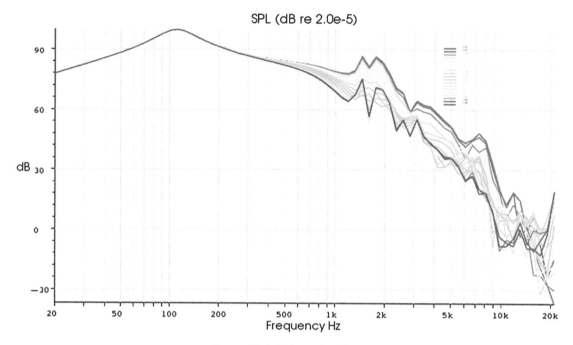

Figure 27.4: Subwoofer SPL.

Figure 27.5: Subwoofer Acoustic Pressure Field.

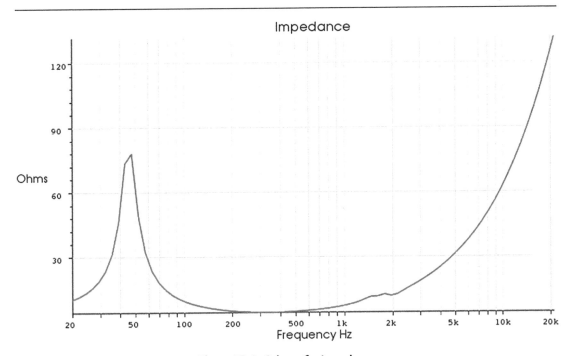

Figure 27.6: Subwoofer Impedance.

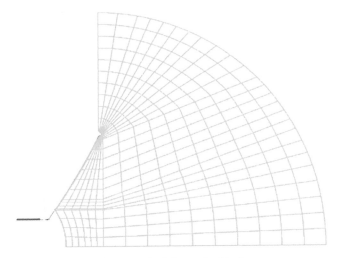

Figure 27.7: Subwoofer Mesh.

Earlier we asked whether this model would give reasonably accurate and, more importantly, reasonably representative results? Our experience of measuring many similar loudspeakers confirms this. It certainly lines up with the theoretical designs presented earlier, but of course it does take them a quite a bit further on, with responses now being modelled up into the breakup region and off-axis. As well as the impedance modelling, this is a very useful tool.

Notes

1. This is not true for compression horns and some other transducer designs where full coupling becomes essential for full modelling accuracy.
2. *Bl* has been manually entered and has no direct relationship to the actual magnet system—just use it as a guide for the response shape and you will be fine.

Reference

[1]   *PAFEC & PafLS.* www.pafec.info/pafec/ (visited on 31/01/2018).

CHAPTER 28

Linkwitz Transform

The Linkwitz transform allows us to change the low-frequency response shape of a closed box loudspeaker system to achieve a more desirable response than the loudspeaker driver plus enclosure would achieve without assistance.

As we have seen earlier, the majority of loudspeaker drivers show a second-order high pass behaviour. Although we have not done so, they can be described by a pair of zeros at the s-plane, and a pair of complex poles at the location of F_s and Q_t.[1]

The Linkwitz transform allows us to place a pair of zeros (F_z, Q_z) on top of the pole pair to exactly compensate their effect. A new pair of poles then may be placed in a more convenient place to achieve the desired response.

See the Siegfried Linkwitz[2] website [1] for mathematical and circuit details on the transformation process.

With our subwoofer we already know that the loudspeaker driver will need assistance to meet our specified output at 20 Hz. The parameters we have so far will not allow us to reach our target. So what can we do to change the parameters whilst still using the same loudspeaker drive unit?

At first sight this seems impossible. However, by applying an active equalisation we can achieve this, albeit with several costs:

- We need an active or DSP system.
- We need an individual power amplifier for the loudspeaker.
- We exchange maximum power or SPL for low frequencies or better response accuracy.
- Additional low-frequency response comes at the expense of extra cone excursion.

We will use WinISD[3] [2], which is the subject of Appendix M on page 335, to demonstrate the use of the Linkwitz Transform in our subwoofer.

Leaving aside the mathematical underpinning of how is this achieved, what are the advantages and disadvantages of this approach?

We start off with our predicted low-frequency response of the driver in our 64 litre enclosure. This shows the slow, early roll-off that's typical of a loudspeaker being used in a small enclosure, with a fall-off in response from a relatively high frequency (often a few hundred Hz). This is shown as figure 28.1, with the final output being $-20\,\text{dB}$ or so at 20 Hz.

Let us look at the simulated SPL at a 1 W input as shown in figure 28.2.

We need to produce an inverse of this curve at low frequencies and use this to change the signal fed into the loudspeaker.

This is the Linkwitz transform, as shown in figure 28.3.

This results in the theoretical response curve as shown in figure 28.4.

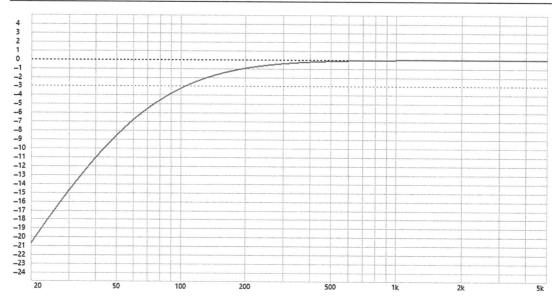

Figure 28.1: Low Frequency Response Simulation.

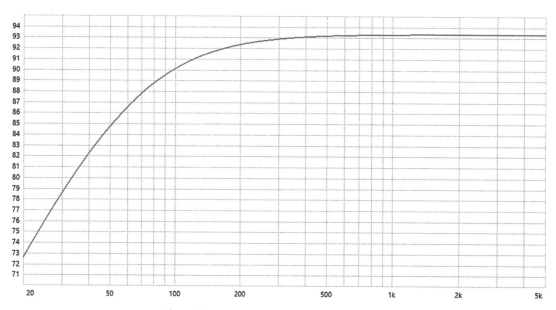

Figure 28.2: Low-Frequency Response at 1 W.

In this case we can see two problems:

*   The response is uneven has ripples at low frequencies.
*   There is an artificially extended response at high frequencies.

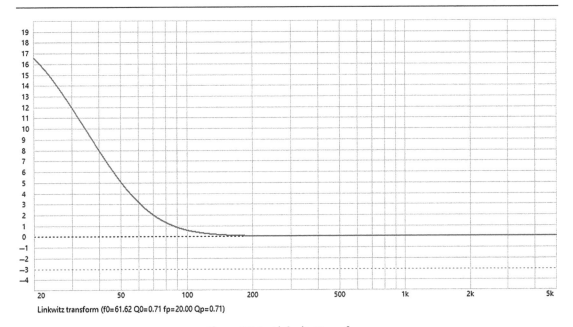

Linkwitz transform (f0=61.62 Q0=0.71 fp=20.00 Qp=0.71)

Figure 28.3: Linkwitz Transform.

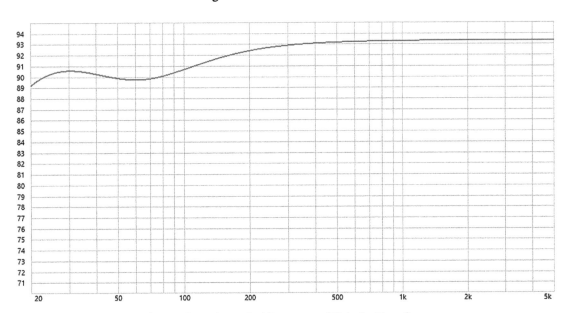

Figure 28.4: Theoretical Response of Linkwitz Transform.

28.1 *Realistic Parameters*

The first is due to our subwoofer not being *ideal*, so the parameters are not as we would theoretically wish. Our first step is to look at the parameters that we have—if we can use measured parameters, so much the better. So let's look at the parameters of our driver unit in our 64-litre closed box as shown in figure 28.5.

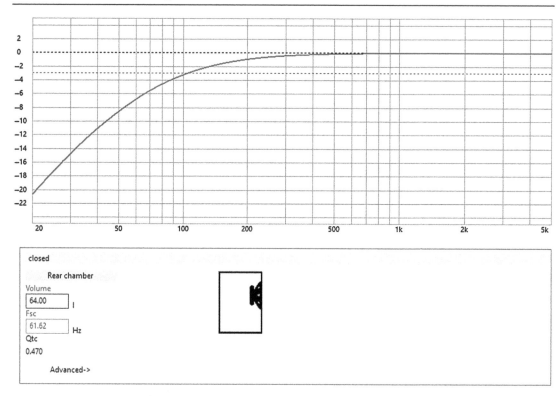

Figure 28.5: Theoretical Response of Driver in 64 Litres.

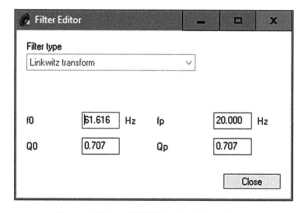

Figure 28.6: Initial Linkwitz Transform.

This shows that our driver and enclosure closed box resonance is 61.62 Hz. That is our first *pole*, and the system $Q_{tc} = 0.470$ is our first *zero*. If we now click on Add Filter *Linkwitz Transform* we see an initial calculation made assuming Q_{tc} of 0.707, as shown in figure 28.6.

So instead of assuming the system Q_{tc} is 0.707, let us input the actual Q_{tc} of 0.470 as shown in figure 28.7.

This results in the theoretical updated Linkwitz transform transfer function shown in figure 28.8.

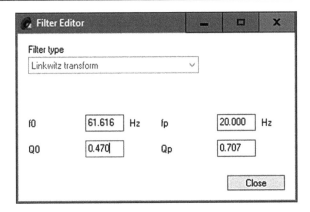

Figure 28.7: Updated Linkwitz Transform.

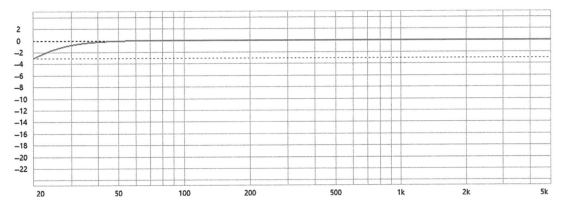

Figure 28.8: Updated Linkwitz Transform Transfer Function.

We can clearly see that the ripples present in figure 28.4 on page 129 have gone, so demonstrating that we can effectively equalise or correct a second-order response curve.

28.2 High Frequency Response

The second problem is a consequence of the lumped parameter model not taking account the roll-off due to the voice coil inductance.[4] This inductance is both high and distinctly nonlinear, ranging from 0.05 H to 0.01 H depending upon the instantaneous displacement. Currently this has only been simulated over the range of ±7 mm.

We will take the zero displacement value of 30 mH. Using the 'Advanced' feature we can simulate the voice coil's impedance to the existing Linkwitz transform; this is shown in figure 28.9.

We now have a response falling off gently at low frequencies but with a significant peak of 12 dB at 75 Hz.

So let us apply a 12 dB notch to this curve; the result is shown in figure 28.10.

This looks quite good with response down to 20 Hz and up to 100 Hz.

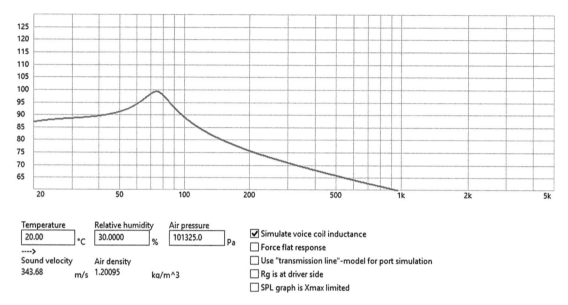

Figure 28.9: More Realistic Low-Frequency Response at 1 W.

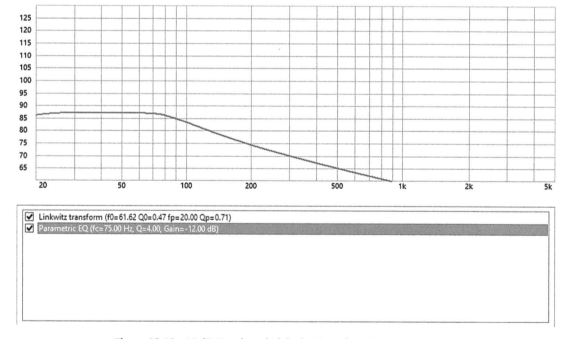

Figure 28.10: 12 dB Notch and Linkwitz Transform Response at 1 W.

28.3 Cone Displacement

The cone displacement is shown in figure 28.11.

However, the SPL is still far below our target of 110 dB SPL at 20 Hz, so let us increase the input to 200 W.

The cone excursion has now increased dramatically to nearly 50 mm. Interestingly, most of the movement is below 30 Hz. This is shown in figure 28.12, and most of the really high movement is below 25 Hz.

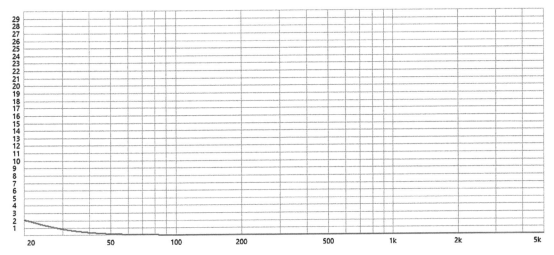

Figure 28.11: Linkwitz Transform Cone Excursion at 1 W.

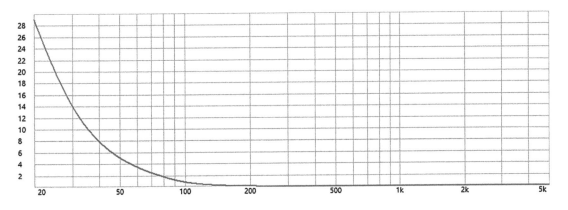

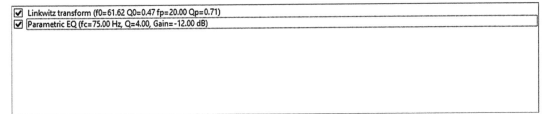

Figure 28.12: Linkwitz Transform Cone Excursion at 200 W.

So let us apply a second-order high pass filter at 25 Hz and see what effect this has. The simulation is shown in figure 28.13.

Cone excursion at 200 W looks much better: less than 15 mm displacement and that only at 25 Hz; otherwise less than 12 mm.

Finally let us examine the final SPL at 200 W as shown in figure 28.14.

Let us just review our target and our results:

- Target 110 dB SPL at 20 Hz.
- Result 104 dB SPL at 20 Hz, peaking to 110 dB at 40 Hz to 80 Hz, 104 dB at 120 Hz.

This has *theoretically* been achieved without floor or wall loading which will increase levels—job done for the moment.

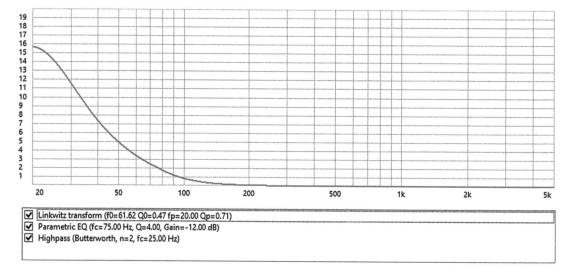

Figure 28.13: Linkwitz Transform with 25 Hz High Pass Filter, Cone Excursion at 200 W.

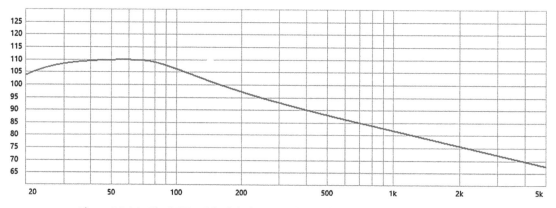

Figure 28.14: Final SPL with Linkwitz Transform with 25 Hz High Pass at 200 W.

Notes

1. Poles are the frequencies and zeros are the damping factors.
2. Siegfried Linkwitz was for many years an engineer and manager with Hewlett Packard.
3. Currently at version 0.7.0.950.
4. See Chapter 7: Small Signal Model on page 37 and the $L_e(x)$ simulation shown in figure 20.4 on page 87.

References

[1]   *Linkwitz Transform.* www.linkwitzlab.com/filters.htm#9 (visited on 01/02/2018).
[2]   *WinISD speaker designing software for Windows.* www.linearteam.org/ (visited on 31/01/2018).

PART VII

Mechanical Design

CHAPTER 29

Visual and Mechanical Design

This chapter will discuss how we start to translate a design into a final loudspeaker, phone, or completed system, going beyond the driver and how it works towards the final result. Please note that we are not particularly visual, and we are only going to outline the process from our perspective.

The visual design or initial concept often starts as a series of sketches or simple 3D models and visual renders. These are used to develop an initial selection, and a choice is made on the overall look.

Often semi-hard foam blocks are used to provide a physical shape. Only when these initial concept stages have been passed is it usual to start CAD drawings. Often the full cabinet enclosure shape will be modelled first, the details of the mechanical parts being filled in later.

We already have the outline of the 3D CAD as we modelled the acoustics using ABEC. True, it lacks detail, but a large part of the mechanical and visual design is filling in the fine detail so that we end up with a functional and hopefully a pleasing visual appearance at the same time, usually at an affordable price.

A 3D master drawing will need to create an overall framework, which of necessity will contain all of the parts, be they drive units, plinths, fixings, grilles, or feet. All these parts will need to be individually specified and drawn but will later be reintegrated to form the whole design.

If using a professional 3D CAD system or Fusion 360, all of these individual parts will constantly be checked to ensure if they fit together, as well as building a full parts list.

One often-asked question is whether there should be a master 3D CAD model containing everything. Our answer is no; we should not create everything in one drawing at one time. In theory and practice, it is more efficient to use a series of sub-assemblies, all of which then fit together. The good news here is that we have already started our master parts list—we just called it a bill of materials (see Chapter 14 on page 62), and this will become the core structure of the assembly or full 3D Model.

As it is essential for a visual model, especially a 3D example, to show all of the proposed detail and render it so that it becomes a true visual representation, and to do this for surface finishes as well as materials, lighting and background will all need to be included.

So the visual or mechanical designer would typically have many tasks. Designing the whole look and feel is probably the most important element. This breaks down into the cabinet and other parts, usually with some interaction between the mechanical and the acoustic or drive unit designer(s), after which the mechanical designer would usually design the chassis in 3D.

The drawings of the chassis would be passed to the drive unit designer who would typically extract a section through the centre line of the chassis, and this would then form the basis of a 2D axisymmetric stack-up drawing.

If a full 3D model is required, the whole chassis drawing may, of course, be used. This leads us neatly into the next section of driver design. In the case of our subwoofer, how should we proceed? Well, we know the outline shape (a 450 mm cube with 25 mm MDF panels).

What else is required?:

1. We need to mount the driver into the cabinet.
2. We probably need to brace the cabinet to stop excessive vibration.
3. We may need to mount an amplifier into the cabinet.
4. We need some feet.
5. We will need to design the following components:
   - Surround.
   - Cone.
   - Dust cap.
   - Chassis.
   - Voice coil.
   - Top plate.
   - Pole plate.
   - Magnet.
   - Terminals, screws, and nuts.

Some more questions:

1. Is the chassis to be flush-mounted into the front? If so, will a 25 mm thick front panel be strong enough?
2. How will the driver be fixed to the cabinet, and will the fixings be visible?
3. Will the cabinet be braced to stop or reduce panel vibration?

If the chassis is to be flush-mounted into the front then 25 mm may be too thin, so a thicker front will be needed. Let's use two panels of 25 mm material, so we can safely rebate the chassis into the outer layer.

Let's decide to design a really clean and contemporary look with no visible fixings. Ideally, we would like a monolithic cube, though we would also like smoothly rounded corners.

It is also essential that the cabinet is braced. We can now start detailed design; let's start with the chassis by importing the motor unit dxf we used to calculate the magnetics with.

If we start with this at the bottom of an axisymmetric model of the full driver, we can work on up until we have completed the whole driver. Starting at the bottom there will be:

1. The back plate (15 mm thick and 105 mm radius) and in the middle of it the pole (54 mm high & 37.25 mm radius).
2. The magnet(s) (40 mm thick and 110 mm radius).
3. The top plate (8 mm thick, 105 mm radius OD, and 38.125 mm radius ID).

We start with the chassis, and we already know that it needs to be quite strong to hold the motor unit which will weigh around 5 kg.

What are the types of chassis and what are some of the implications of using these?

1. Pressed steel.
2. Cast or die-cast alloys.
3. Moulded plastics.
4. Machined steel or alloys.
5. An integral part of the cabinet.
6. Additive manufacturing techniques (MakerBots, stereolithography, sintering).

Let's go through these one at a time:

1. Pressed steel chassis requires either a very high degree of skill or some serious press tools. This used to be the favoured method for high-volume low-cost production.
2. Casting i.e. sand casting is a practical option for small production volumes. However, you are dealing with molten metal, with all the risks that entails. Die-casting requires extensive tooling, so is less suitable for limited volumes.
3. Conventional moulding again requires extensive tooling and so is not suitable for limited volumes.
4. Machined parts would need fairly extensive workshop facilities dependent upon the complexity envisaged—but this is both possible and practicable for one-offs.
5. Many modern consumer loudspeakers are produced as integral parts of a whole—but these usually require injection moulding techniques that require extensive tooling.
6. For the first time, these 'additive manufacturing techniques' allow us to produce components in small quantities, by using the power of 3D modelling and printing, that previously could only be moulded with extensive tooling.

The price to be paid is that tooling the chassis at least will now require a full 3D model of many of the individual parts; fortunately, there are now many software and hardware tools readily available to assist with these tasks.

We develop a potential design for a chassis in Chapter 30 on page 142 and for the cabinet or enclosure in Chapter 31 on page 149.

CHAPTER 30
Chassis

This chapter will consider some of the requirements that a chassis needs to meet, using the same CAD tools as the rest of the mechanical design. The first critical requirement is to support the motor (often the heaviest part of the loudspeaker), so this needs a combination of strength and accuracy.

We need to support the spider (or supporting and centralising parts), fix the lead-out wires to the voice coil and the terminal panel, and design any gaskets or seals required, as well as holding the cone surround assembly while still ensuring sufficient linear movement (X_{max}), all while holding everything together in the correct alignment(s). The final factor is to mount the chassis onto the cabinet without causing acoustic interference.

We well remember, many years ago, having to delay the production of a 15-inch loudspeaker driver, as we had not ensured sufficient clearance under the suspension with the purchased chassis. Then followed a very nervous meeting between a young engineer and the managing director to explain the problem and then agree a solution: 'So, Geoff, will this happen again?' 'No, Roger.' Ever since, we have insisted on using a stackup drawing to ensure that all parts have sufficient movement (X_{max}) available.

How can we achieve all of these requirements? This will depend in part upon what the chassis has to support. For example, if the chassis is for a 3/4-inch (19 mm) tweeter, the supported weight is likely to be quite small. (Furthermore, anything in the acoustic path of a tweeter will tend to have a significant influence, as the wavelengths the tweeter will be reproducing will probably be shorter than the physical size of its chassis.)

However, our subwoofer driver will probably use a large ferrite magnet with steel pole and top plate and a thick, large-diameter backplate. Clearly these are going to weigh a lot, so the chassis designs for these two products are bound to be very different.

We should also consider other relevant factors and components, so let's start with Subwoofer 1 from later in the book. Some rough initial parts and the detail of the modelling are described in Appendix E: Fusion 360 on page 226.

30.1 Stackup Drawing

We will start out with an axisymmetric stackup drawing.[1] We draw it using all of the parts that we currently know, as shown in figure 30.1.

Key dimensions are:

1. The position of the top plate.
2. Clearance around the voice coil.
3. Support for the spider/damper.
4. The support for the surround.

Start by sketching an outline profile of a chassis using just these four reference positions (as shown in figure 30.2).

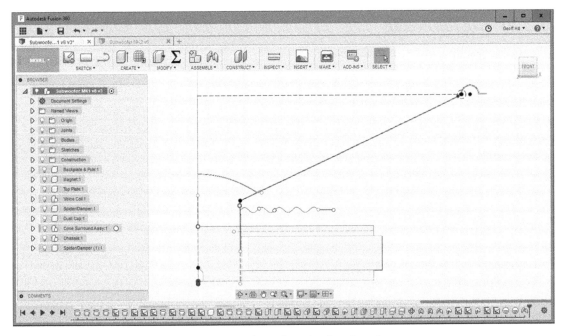

Figure 30.1: Subwoofer 1 Stackup Drawing.

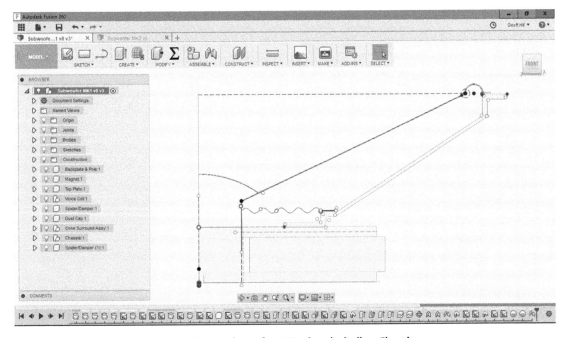

Figure 30.2: Subwoofer 1 Stackup including Chassis.

30.2 Revolve the Profile

Turn off the visibility of all parts except the chassis (as shown in figure 30.3).

Take this profile and revolve using the central Y axis (as shown in figure 30.4).

Obviously this chassis is solid, as shown when we tilt the viewing angle (see figure 30.5).

It is essential to take account of trapped air pressure under parts as these can seriously affect the sound produced by the loudspeaker. We think of air and sound as being somewhat like water: If water is in a container and there is any way out, water will find it. Likewise with sound pressure; many an otherwise fine loudspeaker has been made to sound bad by not taking this into account. In this case the back of the cone would be heavily restricted.[2] So we need to introduce some vents to allow the sound to escape from the loudspeaker, and we may use a profile as shown in figure 30.6.

We then extrude this profile up through the solid chassis;[3] the initial chassis is shown in figure 30.7.

One of the most difficult things to achieve when designing components is ensuring that parts go together correctly; ideally we should self-align the various parts of an assembly. So we will use a series of fixings symmetrically around the central axis (as shown in figure 30.8).

Next we need to attach the chassis to the top plate and enclosure, so we add suitable mounting holes and screws, plus a terminal panel to connect to the voice coil. (The resultant chassis is shown as figure 30.9).

An old trick when using staked top plates and pressed steel chassis was to put a small quantity of soft adhesive between the two parts, as this was very effective in stopping minor buzzes that were difficult to resolve any other way.

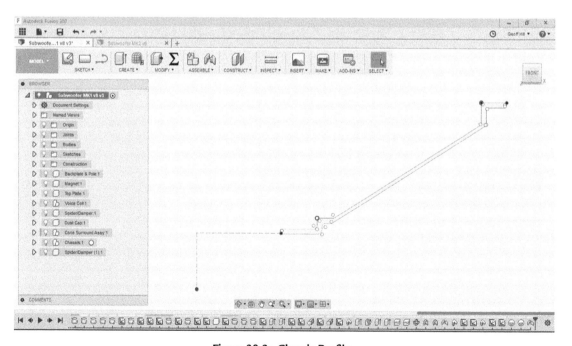

Figure 30.3: Chassis Profile.

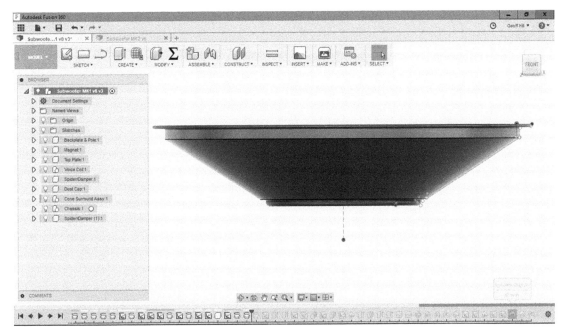

Figure 30.4: Chassis Side View.

Figure 30.5: Chassis Tilted View.

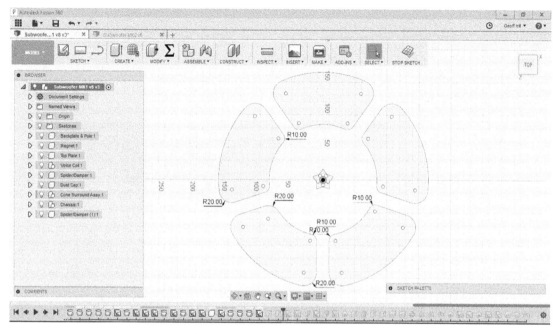

Figure 30.6: Chassis Bottom Profile.

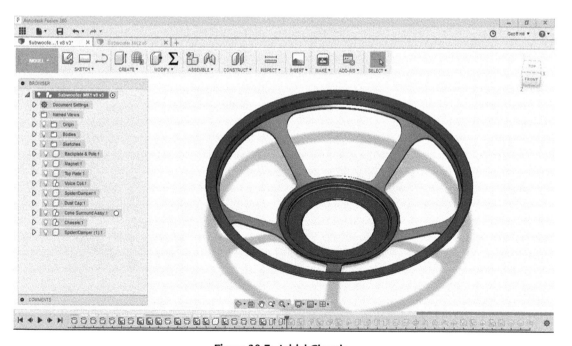

Figure 30.7: Initial Chassis.

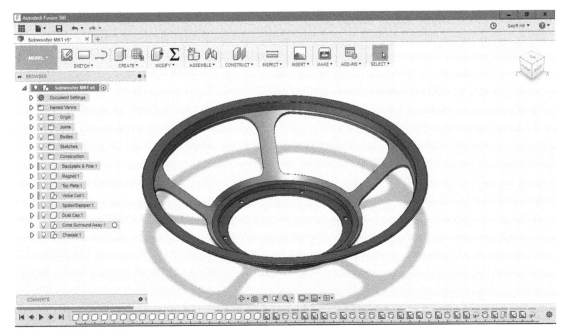

Figure 30.8: Chassis.

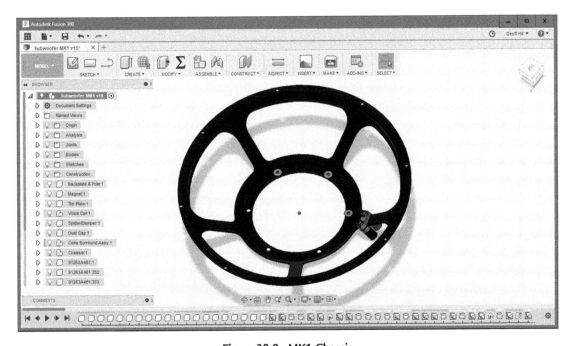

Figure 30.9: MK1 Chassis.

There are many other things to consider, including:

- Visual finish.
- Mechanical strength.
- Physical protection.
- Mounting to the cabinet or enclosure.

Plus many others, but we will leave this chapter here.

Notes

1. A stackup drawing shows the physical relationships, sizes, and alignments between all of the parts.
2. Which might make a good waterproof loudspeaker.
3. In 3D CAD modelling, the term 'extrude' is often used to denote either extruding (adding material) or else cutting (removing material).

CHAPTER 31

Cabinet

This chapter discusses the cabinet or enclosure, as this often forms a large part of the mechanical structure. (Indeed, in a mobile phone or a tablet it may well be the core of the physical structure). In a complete loudspeaker system, the cabinet or enclosure is often the single largest item, and it certainly has a significant effect on the performance of the whole system.

We considered earlier some of the considerations of an enclosure on the low-frequency response. However, it is important to say that we have only considered a very small range of system types, and only at the lowest frequencies.

What do we need a cabinet for? The fundamental reason comes back to the old nemesis of efficiency. Conventional loudspeakers are horrendously inefficient, and part of this inefficiency stems from the fact that they are often considerably smaller than the wavelengths they are reproducing.

This is especially true as the sound is usually produced by both sides of a drive unit, and each is out of phase with the other, so the outputs will readily cancel out at low frequencies.

A major function of the cabinet is to prevent this occurring, and it is also often called upon to control the response at various frequencies. There are various ways that it does this:

- Open baffle.
- Closed enclosure (or embedding in a wall).
- Vented, ported enclosure, or transmission line.
- Bandpass enclosure.
- Horn loading or waveguides.
- Loudspeaker arrays.

All of these will significantly change the performance at low frequencies, and the reader is directed to many papers and books published about these individual methods and techniques.

However, with all of these methods the cabinet or enclosure plays a critical part at higher frequencies as well, because of driver interaction and diffraction. This is especially so with horn, waveguide, and array types of systems, where enclosure design takes on a fundamental role in setting or controlling aspects of the frequency response.

In a commercial setting, the necessary task of cabinet design would be happening in parallel with other procedures. Here we need to match the driver(s) with the chosen cabinet type and the desired materials and finishes along with producing the necessary drawings.

Ultimately there are two main schools of loudspeaker design: Cabinets should either be as light and strong as possible, so as not to store energy, or cabinets should be strong and heavy so that they are unaffected by the movements produced by the drive units.

The advent of modern high power amplification has led to a marked reduction in efficiency, as we noted in the chapters on Thiele/Small parameters. However, this reduction in efficiency is not without performance costs.

We covered the basics of those due to temperature and mechanical movement in the large signal parameters chapters. These can be analysed during the design of loudspeaker drivers, by the various tools that we have discussed.

However, there are a few areas where the cabinet and cabinet shape can have a very strong effect. One is that of baffling and diffraction. Another is where the cabinet shape allows (or encourages) the formation of standing waves (although some designs like 1/4-wave resonators are based upon using standing waves). A third major problem is that the surface area of the enclosure is usually much larger than that of the radiating area of the loudspeaker driver(s). Thus, the surface area of the cabinet or the enclosure can easily radiate significant levels of sound. Furthermore, this sound is rarely as well reproduced as that from the drivers, and it can easily interfere with the desired sound.

31.1 Subwoofer Cabinet

We decided earlier a rough outline for our subwoofer enclosure design—we look further into the acoustic modelling later in Appendix C on page 189.

What do we currently know? Size $450 \times 450 \times 450$ mm, closed enclosure, 25 mm enclosure walls, chassis as discussed in Chapter 30 on page 142; and our overall specification was outlined in Chapter 16 on page 71 as follows: a loudspeaker driver with a 348 mm effective diameter; a 500 W rms amplifier; and a system that can produce 110 dB SPL at 20 Hz with distortion below 10%. (We do not need to examine this deeper at the moment.)

But for the cabinet design we need the mechanical dimensions and also the knowledge that the subwoofer will be producing very high pressures at very low frequencies. Although the panels themselves are relatively small and panel modes are probably not significant over the operating range, the enclosure still needs to be extremely strong to withstand the specified pressures and levels without physically destroying itself.

To do this we need to brace the cabinet and fix in place the front, back, and bracing components strongly, along with ensuring free air flow, as shown in figure 31.1.

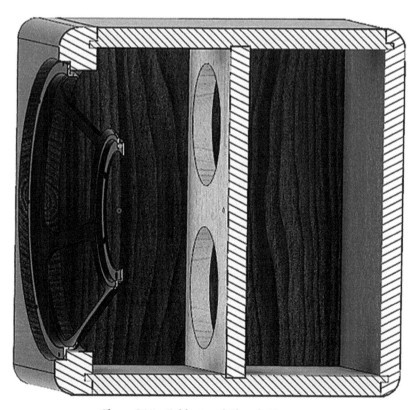

Figure 31.1: Cabinet and Chassis Cutaway.

PART VIII

Measuring a Loudspeaker

CHAPTER 32

Acoustical Measurements

This chapter will begin by examining why acoustical measurements are different from electrical or electronic ones and discussing how to make reliable, consistent, and accurate acoustic measurements. We look at conventional anechoic chambers before pinning down the real underlying requirements for measuring loudspeakers. We consider the accurate measurements of loudspeaker drive units together with mounting baffles, and the numerous ways this could (and often does) go wrong.

Most electrical or electronic measurements are made with instrumentation that makes little or no changes to the device being measured, and these are usually linear in amplitude and frequency (over their operating range). They are also usually taken at specific points and are only influenced by the environment to a very small (often negligible) degree. Finally, most electrical or electronic measurements are highly repeatable and generally have relatively good linearity and low distortion.

However, acoustical phenomena are usually measured using microphones that convert the acoustic wavefront into an electronic signal. Unfortunately, if a microphone is large compared to some of the wavelengths it is converting, there's a possibility of acoustic interference, especially so at high frequencies, and this is often a reason why a curve looks 'furry'.

Then the acoustic measurements are or can be heavily influenced by the environment, due to reflections, absorption, and diffraction. Measurements are typically made at just a few points in a three-dimensional environment, though there is increasing demand for 'balloon' type measurements that sample the acoustic field over a full 360° sphere [4]. (However, this is the exception rather than the rule.)

Often measurements change dramatically with relatively small changes in position, and some microphone types have very nonlinear sensitivity that depends upon the distance and position between the acoustic source and the microphone. Cardioid microphones for example. You may wonder that we are able to make any measurements at all under these circumstances. And a case can be made for not measuring loudspeakers in a listening room because of the inevitable reflections, absorption, comb filtering, and other effects. This is even taking into consideration the advances in measurement equipment theory from Heyser [5], Fincham and Berman [2], Vanderkooy and Rife [8], Farina [3], and Klippel [6].

Let's make it plain at this point that in this chapter we are talking purely of making measurements to help us characterise and understand the design and performance of loudspeakers. Specifically, we are not concerned with the recording of speech or music for later reproduction.

Over the years that we have been involved with the design, test, and measurement of loudspeakers, one thing above all other stands out as of primary importance. When it comes to making accurate and repeatable measurements on loudspeakers, it is the environment, and specifically the setting up of simple systems to reduce inaccuracies, that really matters the most.

We have worked in dozens of anechoic chambers over the years, from those one had difficulty fitting inside to those with room for a heavy goods vehicle and its 40-foot trailer. One at a long-gone company was even so badly designed that if you went into it and closed the outer door and then the inner door you would literally be locked inside with no way out! So it was company policy that if you needed or wished to make any measurements, at least two people were required; oh, yes, it also had a typically bouncy mesh floor which made accurate set-up impossible. An enormous steel beam and floor mesh splitting the chamber at floor level totally mucking up the semi-anechoic measurements as well.

All of these were purported to be anechoic chambers, designed so that the attenuation versus distance conformed to square law behaviour over as wide a physical range as possible. All the chambers were ideally designed to work reliably to as low a frequency as possible, given the cost and space constraints of the particular company or faculty. As we saw earlier, another factor in the design of most anechoic chambers is the pursuit of the lowest possible noise floor. Most people unfortunately think that a low noise floor alone will ensure suitability for the measurement of loudspeakers. We must disagree. However, many such anechoic chambers are not well suited for a range of reasons.

1.   They are usually not able to hold the loudspeaker and the microphone in a fixed geometrical relationship.
2.   They (or the way they are used) introduce extra uncontrolled reflections from objects within the anechoic chamber.
3.   They often introduce diffraction effects when measuring drive units of different sizes.
4.   They were usually designed to minimise the ingress of external noise to the point where other operational considerations were ignored or overlooked.

Most people think of anechoic chambers as rooms without echoes, but minimising external noise is actually a minor point. Yes, any noise will affect the accuracy of a measurement, but in reality, we are surrounded by noise at differing levels all the time. So our brains are rather well equipped to filter out random background noise in particular (wind and rain, for example) and to notice certain things over others. The real question is rather: Is noise affecting any given measurement significantly or not? And our modern tools—FFT, etc.—can and do mirror this capability.

Our measurement analysis tools have moved well beyond the crude 1/3-octave analysis where only energy could be reliably detected to the point that we can easily examine in detail the impulse or step response of a system and work effectively in the time domain.

Just having a low measurement noise floor is not the 'be all and end all'; we need to focus on those measurements that are relevant. This is especially true now that modern software measurement systems have many ways of effectively 'seeing' through noise to get at and analyse even very low-level signals.

Since that badly designed anechoic chamber, we have been fortunate enough to design many other examples used specifically for loudspeaker design, test, and verification. In these designs we have tried to design out any of the problems suffered in the past.

This is particularly relevant as the investment in time, materials, and physical space is not to be taken lightly: For most companies, it's a one-off investment that must last years. Using many chambers over the years provides an insight into the design and use of an anechoic chamber for measuring loudspeakers, and there are often problems in ensuring consistency of the final measurements. Other problems include ease of access, difficulty or inability to change set-up between the different measurement requirements, and the clash between the requirements of a production or quality control (QC) environment and the requirements for R&D and Design.

32.1 Conventional Anechoic Chambers and Wedges

In a classic paper by H.P. Sleeper, Jr. and L.L. Beranek [1], 'The Harvard Anechoic Chamber', CIR-51, Electro-Acoustic Laboratory, Harvard University, 1945, a wedge design geometry was established that is still in widespread use today.

Typically, these have tended to be full anechoic chambers in the loudspeaker industry. The preference is towards the semi-anechoic chamber in the automotive industry. Chambers are mostly designed with grids, meshes, or wires to support the devices under test, so usually anechoic chambers are either built into pits in the ground or their entrance(s) are via raised stairs or ramps, so access tends to be compromised.

Let's examine what a typical anechoic chamber is designed to achieve and the relevance for loudspeaker measurements:

1. Free field sound dropping at 6 dB per octave at all frequencies.
2. Free from reflections.
3. Isolation of external influences upon the device being measured.
4. Measurements able to be conducted anywhere within the volume.

Okay, so let us take these in turn:

1. Good if it works, as this means that no additional energy is being returned to the device being measured—but it can never be achieved at all frequencies.
2. Great in theory, but in practice if they are only 20 dB down they could be very hard to find and often come from grids and floors rather than from where you might expect.
3. Again, great in theory, but a lot of money is spent ensuring a low noise floor; currently the best are in the region −10 dB SPL to −13 dB SPL. While mostly we are measuring loudspeakers at 70, 80, 90, or 100 dB SPL, and using tools that can easily see down into the noise more than −100 dB!
4. Great! We can measure anywhere, which means in practice that different people will measure differently every time.

Looked at it this way, it's hardly a recipe for successful measurements.

Many of the problems with conventional chambers and current recommendations (IEC 268-5 and JIS C5531) are discussed by Alan S. Phillips' paper 'Measuring the True Acoustical Response of Loudspeakers' [7].

The IEC Standard 265-5 recommends the use of a large baffle for drive unit measurements; however, how to mount the driver is not clear. Whilst the JIS Standard C5531 recommends using a large 640-litre enclosure, both specify using their individual recommendations within an existing anechoic chamber but do not address the other problems and inconsistencies introduced by using such a methodology.

32.2 Underlying Design Requirements

What do we need for an anechoic chamber to be successfully used for the production, QC, and R&D of loudspeakers? Obviously it must be able to cope with a range of potentially conflicting requirements:

1. Ease of set-up (and strip-down).
2. Repeatable, consistent, and accurate measurements.
3. Full anechoic measurements at 1 m (ideally 2 m) for cabinet systems.[1]
4. Drive units measured in a baffle.
5. Capable of polar measurements on cabinet systems.
6. Quick changeover for production cabinets.

It may seem strange that we put ease of set-up and strip-down at the top of the priority list, but this is based upon the experience of many chambers that have been difficult to set up or change. Most people try to speed the process up by taking 'short cuts', and almost inevitably these compromise the measurements they are undertaking.

The difficulty in changing between set-ups means that in practice a changeover is never actually completed, so that the chamber becomes stuck in a sort of limbo, capable of neither full anechoic work nor accurate semi-anechoic work.

By ensuring that it's easy to set-up and strip-down, it becomes easier to ensure repeatable, consistent, and accurate measurements; everything else can then follow in turn.

32.3 Accurate Measurements of Drive Units

We agree with Alan Phillips's comments on the use of a baffle for drive unit measurements. However, we disagree with the notion that drive units should always be mounted from the front, on the basis that is how they are designed to be used, for the following reasons:

- This dramatically increases the risk of damaging the driver.
- It requires access to the inside of the chamber.
- It increases the risk of disturbing the microphone position.

A more practical method is often to mount the drive unit from the rear, minimising any problems raised by design and calibration. We have found that one of the best and most reliable methods for measuring drive units is through the use of interchangeable sub-baffles. These are then mounted into the main baffle in the chamber. This then separates the mounting of any individual driver from the design of the chamber as a whole.

Focusing on the sub-baffle, its main requirements are:

- To provide a simple and reliable method of changing between drivers.
- To provide a precisely repeatable and secure mounting for a driver.
- To change the acoustic loading of a driver as little as possible.
- To accommodate a large physical size range of drive units.

We therefore recommend the use of a sub-baffle/main baffle construction for loudspeaker drive unit measurements.

The driver is clamped from the rear into the sub-baffle and the latter is clamped from the rear onto the main baffle. The clearances between the sub-baffle and the driver and the main baffle are both within ±0.1 mm, between the driver's mounting face and the face of the sub-baffle, with the face of the sub-baffle being flat to the face of the main baffle (±0.1 mm).

See figure 32.1.

Given the baffle, we should now turn our attention to the microphone and the environment.

We really cannot help but repeat here that when someone (usually a boss or QC manager) has complained about a lack of accuracy in the data reported to a customer, nearly every time it has come down to a lack of precision in the set-up, such as the wrong levels being used or something equally simple.

Figure 32.1: Measurement Baffle.

This has often been made worse by the tendency of people to set up individually from scratch every time, so although the same measurement may have been made, it is often the case that:

1. A different microphone and/or pre-amplifier was used. (And exactly which calibration curve was used, and what was the reference level?).
2. It was fed from a different power amplifier, the level being set using the 'un-calibrated' multimeter kept in the QC department cupboard that everyone had forgotten about.
3. It was measured in an unknown position in the anechoic chamber.
4. The loudspeaker was placed on an 'orange box' that was just the right size for that loudspeaker.
5. The microphone was a 1/2-inch free field type (or was it a 1/4-inch pressure device?).
6. We know it was measured on the Audio Precision gear (or was it the CLIO or WinMLS?).
7. Oh no! We recorded the waveform by sound card and then post-processed the data in MATLAB/SciLab.

Before plotting the data in Excel/Open Office, the report is written in Word/Open Office or LaTeX. We are joking here of course, but we are equally sure you can see that there are an awful lot of potential mis-steps in this process.

What to do? First and most important is the need to define and keep exactly the same set-up(s) for all measurements. In our view it is essential always to use exactly the same microphones and keep them mounted in exactly the same way for every measurement.

Then we would strongly recommend deciding upon one test environment and sticking to it. For loudspeaker measurements, consistency is (in our opinion) more important than absolute noise levels or even reflections (unless the latter are badly out of control).

At a stroke you have eliminated most of the worst causes of inaccurate loudspeaker measurement. So let us now look at the environment and microphone position. Here we have a dichotomy.

For most loudspeaker systems, we are interested in the performance of the raw system and wish to take it out of the listening room. So in most cases we are essentially after the 'free field' response.

Most loudspeaker drive units will be used in some form of baffle or enclosure, so it makes sense to use a baffle, as discussed earlier. To our mind both the IEC and JIS approaches fudge the issues, by putting either a baffle (IEC) or a box (JIS) within an existing anechoic chamber.

Inevitably, this compromises the anechoic chamber even more seriously, as in practice both methods are so large that they are inevitably 'tucked away' in a corner of the chamber at best—or more usually just left where the last measurement was taken, thus forcing further loudspeaker system measurements into undesirable positions within the chamber.

How to overcome this fundamental problem? To our mind it is essential to move away from an anechoic chamber (room) where we can measure any way we like, to one that measures a loudspeaker system or drive unit in a defined acoustical environment, and one that will not change from measurement to measurement.

This means that we require two acoustical environments as a minimum, though it is possible to combine them in one structure given some time and effort.

32.4 Loudspeaker Measurement—Anechoic Chamber

Figure 32.2 is based upon several chambers that we designed and installed for companies in the U.K. and China.

Looking from the top with the wedges removed we see the floor plan shown in figure 32.3.

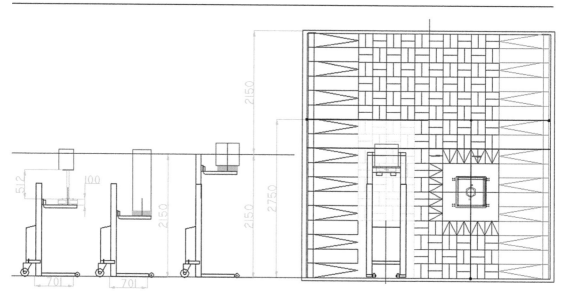

Figure 32.2: Anechoic Chamber Inside Front View.

In this chamber we have thrown away much of the 'rule book':

1.   By dispensing with the mesh floor so beloved of most people (along with its problems).
2.   Literally hanging the microphones down from cross wires at fixed positions 1 m and 2 m for the complete speaker system, and 0.5 m for the driver.
3.   Bringing the loudspeaker under test to a defined X, Y, and Z coordinate within the chamber.
4.   Using a modified lifter and stacker to position the loudspeaker at the correct measurement position (around halfway up the chamber height).

The advantages of this approach are many, including:

*   Stable and defined measurement position (which allows reliable loudspeaker/chamber calibrations to be made).
*   Easy and quick access to the loudspeaker under test.
*   Reduction in manual handling.
*   Fixed permanent microphone(s) which can easily be calibrated for maximum accuracy.

Looking to the right of the loudspeaker under test we can see the separated 'baffle' area. Although not ideal from an absolute perspective, particularly if full polar measurements are required, this arrangement still allows a rapid changeover between the different operating modes.

If full polar measurements are required, the entire baffle section is designed to be removed and replaced with a full set of wedges, restoring the entire chamber to full anechoic mode. If such work would need to be done on a regular basis then it would probably be best to separate out the driver measurement into its own test box.

However, in baffle mode we have a separately defined and controlled acoustic environment, comprising a completely flat acoustically reflective baffle in excess of 1 m² with a microphone at a fixed distance of 0.5 m from the nominal centre position. Half a metre is necessarily a compromise, but is near enough to give a welcome +6 dB on the measured sensitivity (easily calibrated out), and reduced strength of the remaining reflections. However, it could not be considered to be in the far field for larger drive units above, say, 12-inch (300 mm) diameter. (This is unlikely to be a problem, as the majority of modern loudspeaker drivers are smaller than this.)

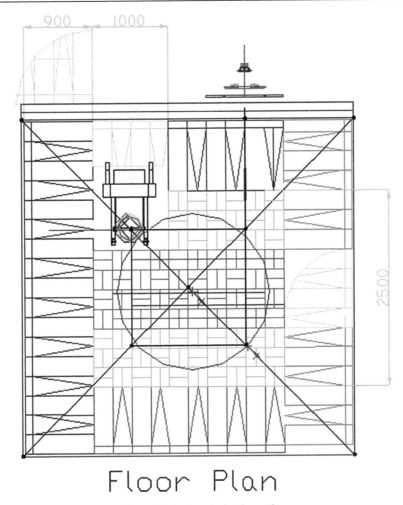

Floor Plan

Figure 32.3: Anechoic Floor Plan.

Note

1. Even 10 m might be insufficient for those working in the public address field.

References

[1]   Leo L. Beranek and Harvey P. Sleeper Jr. "The design and construction of anechoic sound chambers". In: *J. Acoust. Soc. Am.* 18.1 (1946), pp. 140–150.

[2]   J. Michael Berman and Laurie R. Fincham. "The Application of Digital Techniques to the Measurement of Loudspeakers". In: *J. Audio Eng. Soc* 25.6 (1977), pp. 370–384. www.aes.org/e-lib/browse.cfm?elib=3366.

[3]   Angelo Farina. "Simultaneous Measurement of Impulse Response and Distortion with a Swept-Sine Technique". In: *Audio Engineering Society Convention 108*. Feb. 2000. www.aes.org/e-lib/browse.cfm?elib=10211.

[4]   David W. Gunness. "Loudspeaker Directional Response Measurement". In: *Audio Engineering Society Convention 89*. Sept. 1990. www.aes.org/e-lib/browse.cfm?elib=5706.

[5]   Richard C. Heyser. *An anthology of the works of Richard C. Heyser on measurement, analysis, and perception*. Audio Engineering Society, 1988. www.aes.org/publications/anthologies/downloads/jaes%5C_time-delay-spectrometry-anthology.pdf (visited on 04/02/2018).

[6]   Wolfgang Klippel. "Loudspeaker Nonlinearities – Causes, Parameters, Symptoms". In: *Audio Engineering Society Convention 119*. Oct. 2005. www.aes.org/e-lib/browse.cfm?elib=13346.

[7]   Alan S. Phillips. *Measuring the True Acoustical Response of Loudspeakers*. Tech. rep. SAE Technical Paper, 2004.

[8]   Douglas D. Rife and John Vanderkooy. "Transfer-Function Measurement with Maximum-Length Sequences". In: *J. Audio Eng. Soc* 37.6 (1989), pp. 419–444. www.aes.org/e-lib/browse.cfm?elib=6086.

CHAPTER 33

A Universal Loudspeaker Driver Test System

This chapter discusses the ideas behind a universal loudspeaker driver test system, together with the rationale behind AES's X-223 project [1]—what the alternatives are and how we could verify and confirm performance. We then introduce the tetrahedral test chambers.

Okay, let's assume that you either currently or would like to design, manufacture, or test and measure loudspeaker drivers. So what's the problem? Go out there, spend the money, buy the kit and build your system(s). It's simple, isn't it?

Well, actually, it's not that simple. Most people are using ad hoc halfway-house solutions—usually or often designed by companies or people who have no real experience in the measurement of loudspeakers and what is actually required, or who are using older chambers put together by engineers decades ago.

The AES have summed up the problem very neatly in the rationale behind Project X-223:

> In the manufacture of loudspeaker drivers, a frequent final test places each driver into a relatively compact test chamber, measures its acoustic response to a set of test signals, and compares them with stored values derived from a reference sample. This does not provide the full performance measurements that would require a full-sized anechoic chamber, however it can identify deviations from the expected performance set by the reference.
>
> Such test chambers are typically built only to meet local needs. As a result, the figures derived from one cannot be compared with figures derived from another.

Also, where do you get the kit from? Every vendor is pushing their individual solution and to be quite honest we have our own ideas, preferences, and experiences which guide us as well.

So you would really like to know what is going on before you:

1. Commit to production.
2. Send products to your customers.
3. Build into your product(s).

You'll probably agree with these, and surely there are loads of systems that will do these things, aren't there? Well, yes and no. The problem is that most of these solutions concentrate their efforts upon meeting frankly meaningless technical specifications: the easy low-hanging fruit of the software and hardware, rather than the critically important issues.

The important issues are acoustical and mechanical. These together underpin the measurement tolerance that the measurement system can achieve—what the automotive industry calls *Gage R&R*, more fully described as gage repeatability and reproducibility.

Let's go back to basics for a moment. How do you make any reliable and repeatable measurement? Quite obviously you need to measure in a repeatable manner.

For loudspeaker drive units there are many standards. Unfortunately, a lot of them are old and really not up to the job—oh yes, they will work, but only if you really take a lot of care in setting up and so forth.

The first of these standards is the so-called IEC baffle, which actually comes in four sizes, from a nominal 8-inch driver to a nominal 15-inch. What happens if you happen to be measuring 8 × 6 mm microspeakers? Chinese colleagues have recently proposed a fifth smaller baffle for these.

The selected IEC baffle needs to be set up inside an anechoic chamber (quite often the size and cost of a small house). The loudspeaker drive unit to be measured is set up in the baffle, but the fixing conditions are not specified.

A measurement microphone (again unspecified) is set up in the far field of the loudspeaker under test and the resultant measurements referred back to the sensitivity at 1 m.

The second standard is the JIS test cabinet. Basically this is a 640-litre closed cabinet which mounts and holds the loudspeaker under test. A similar measurement is again undertaken inside an anechoic chamber, and with the same problems as the first method.

The problem with both of these approaches is that the acoustic environment is inherently uncontrolled, even with an anechoic chamber. In practice, it is incredibly difficult to keep things exactly the same and uncluttered. A better approach is to say that we need an acoustical environment with a fixed geometrical relationship. So let's build our loudspeaker test system inside out:

- Let us fix our microphone position.
- Let us have easily interchangeable measurement baffles.
- Let these be specific for the individual loudspeaker drive units.
- Let us make it really simple and as compact as we can.
- Let us reduce to a minimum the set-up and changeover time.
- Let us provide a simple amplifier with automated set-up and calibration if possible.

How to do all this? Most people making a test and measurement box do just that—stick a microphone inside a conventional cabinet and go from there.

Unfortunately this is not ideal, as symmetry encourages standing waves to build up. These occur where major fractions or complete wavelengths of a frequency fit an exact enclosure dimension. Obviously, a symmetrical design is more prone to this problem (and most loudspeakers are depressingly symmetrical). So how can we get round these problems?

Loudspeaker driver measurements are well known (in the industry) as a source of potential trouble and conflict: it has been this way certainly since the 1970s when the author started working in this area. Arguments have ranged back and forth over this time (and probably before) without being truly resolved.

So how should we make our measurements? Many of the most consistent measurements that we have made have had either loudspeaker or microphone situated relatively near a corner. This might go against the advice of experts,
but consistency is the higher priority over absolute accuracy. And we believe that it should be your high priority as well.

Why? Measuring from this position gives us a known and relatively stable fixed acoustic gain or transfer curve which is easily removed or corrected out of the final results.

When yet another company needed a loudspeaker measurement solution in 2013, we decided to tackle this problem once and for all. We realised that some of the best loudspeaker measurements had been made with the loudspeaker firing into a corner, and this was later developed into the *tetrahedral test chamber* (TTC).

The first TTC900 model was demonstrated to the AES Conference on Loudspeakers and Headphones in Helsinki, Finland in August 2013, and the TTC350 was then developed from Helsinki feedback.

The concept was formally raised to the awareness of the AES as a whole at the 135th Convention in New York in 2013 [5]. At this convention, we were asked to contribute a draft submission to the X-223 Project on comparative loudspeaker measurement chambers.

We presented a paper to ISEAT 2013 in Shenzhen, China in both English and Chinese and this is available from Hill Acoustics [3]. In March 2014, we presented the TTC to the U.K. Section of the AES in London; a recording of this presentation is available on the AESUK website [2].

A couple of months later we presented an engineering brief at the AES convention in Berlin and to the AES Standards Committee (SC 04-03) [4], and a decision was made to issue an information document as the output of AES X-223 on the subject of 'Loudspeaker driver comparison chambers'.

An initial draft of an AES X-223 document was presented to the AES SC-04-03 committee at the 137th Convention in Los Angeles, but revisions were requested to the structure of the document to bring it more line with an official AES document. The revisions requested were made and the document was presented to the AES SC-04-03 committee at the 139th Convention in New York.

Measurements have now been made with five sizes of chambers from 1-inch tweeters up to 15-inch drivers and full loudspeaker systems, using a wide range of modern FFT-based analysers, all of which give consistent and accurate loudspeaker measurements that have been confirmed independently.

Today, different size tetrahedral test chambers are available for a wide range of drivers and loudspeakers. Used in multiple locations, they are providing consistently stable and accurate loudspeaker measurements in locations throughout the world.

The next chapter is a white paper that examines the theoretical and practical results provided by these tetrahedral test chambers.

References

[1]   www.aes.org/standards/meetings/init-projects/aes-x223-init.cfm (visited on 06/02/2018).
[2]   www.aes-media.org/sections/uk/meetings/AESUK_lecture_1403.mp3 (visited on 06/02/2018).
[3]   *Hill Acoustics.* www.hillacoustics.com (visited on 28/02/2018).
[4]   Geoff Hill. "Comparative Results between Loudspeaker Measurements Using a Tetrahedral Enclosure and Other Methods". In: *Audio Engineering Society Convention 136.* Apr. 2014. www.aes.org/e-lib/browse.cfm?elib=17152.
[5]   Geoff Hill. "Consistently Stable Loudspeaker Measurements Using a Tetrahedral Enclosure". In: *Audio Engineering Society Convention 135.* Oct. 2013. www.aes.org/e-lib/browse.cfm?elib=16958.

CHAPTER 34

Tetrahedral Test Chamber—White Paper

In this chapter we present a copy of a white paper documenting the performance of a SEAS H1207 bass/mid driver as measured by an IEC baffle configured as ground plane and measurements conducted in TTC350 and TTC750 test chambers.

34.1 Aim

The aim of this report is to make comparisons between the loudspeaker measurements made using a traditional IEC baffle (configured as a ground plane) and those made using a tetrahedral test chamber (TTC) and to review the suitability of using TTCs for driver measurements throughout the supply chain.

34.2 Summary

This report shows that from both a theoretical and measurement points of view, TTCs are capable of giving accurate measurements within a relatively small physical size. It also demonstrates a vastly reduced cost in manpower, both in time and skill required to achieve a repeatable result, alongside reduced capital costs compared with traditional methods.

The measurements are shown to correlate very well with those from an IEC baffle. However, they are of higher resolution, not showing the rough responses of a typical IEC baffle measurements.

The use of a TTC will dramatically simplify setting measurement tolerances on loudspeakers by eliminating practically all of the modal effects of current test enclosures. When used throughout the supply chain, the TTCs give a consistency of acoustic measurement that may lead to lower costs and higher quality products by eliminating the current situation of differing acoustic measurement set-ups at different stages.

34.3 Background

Loudspeaker measurements have long been a challenging area, in spite of the theoretical simplicity of such measurements. The earliest loudspeakers tended to be used on a baffle of finite size.

So the earliest measurements probably just followed this as a matter of convenience. A baffle will act to separate the frontal radiation from the rear radiation and this prevents cancellation above a nominal cut-off frequency; as is well known, the size of anything other than an infinite baffle will have significant diffraction of sound from the edges.

Towards the end of World War II there was a requirement to measure the sound from military vehicles and the Harvard Anechoic Chamber was constructed as a result [3]. Most anechoic chambers are primarily designed to be a low noise environment as the military requirement was to detect noise from military vehicles.

After World War II, an expanding economy required more loudspeakers, and organisations such as ISO, AES, and JIS were involved in loudspeaker measurements, notable examples being ISO 268-5, latterly updated to IEC 60285-5, and JIS. These introduced standardized measurement baffles and recommended test boxes; however, these standards required the use of an anechoic chamber to provide the acoustical test environment.

Many of the problems with this methodology were discussed in Alan S. Phillips's paper 'The True Acoustical Response of Loudspeakers' [7] in 2004, with Philips recommending the use of a hemi-anechoic chamber as the primary acoustical measurement environment.

Meanwhile, Richard Small's publication of 'Simplified Loudspeaker Measurements at Low Frequencies' [8] and Don Keele's publication of 'Low-Frequency Loudspeaker Assessment by Nearfield Sound-Pressure Measurement' [6] together provide a practical method of making reliably consistent measurements.

34.4 Tetrahedral Test Chamber

With the TTCs we combine the low-frequency measurement methods outlined by Small and Keele, introducing these into an environment that reduces the effects of internal standing waves by mechanical design rather than passive absorption.

At the very lowest frequencies we *cannot* physically absorb the wavelengths at very low frequencies.[1] So they are long compared to the dimensions of practical chambers; we accept this and calibrate the chambers' response instead at low frequencies, whilst providing effective absorption at higher frequencies where it is most effective.

This can be seen clearly in figure 34.1.

This clearly shows the effective absorption falling off at lower frequencies, and more so for thinner materials, and this shows why conventional anechoic chambers need so much absorption material to absorb (or try to absorb) all frequencies equally.

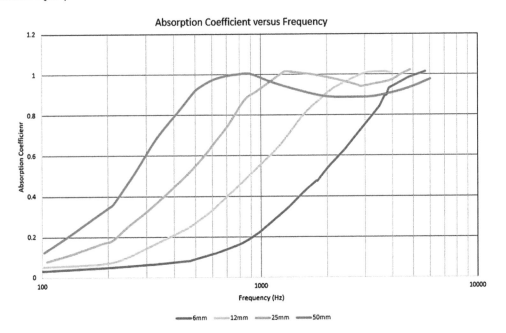

Figure 34.1: Typical Acoustic Absorption Curve.

The last reason we can and indeed should move away from an anechoic environment is that we are usually making *loud*-speakers: even an extremely inefficient one with 0.01% efficiency will easily produce 70 dB SPL re 20 μPa in its passband from 1 W, and most loudspeakers will produce considerably higher levels.

So we should not be so concerned with measurements at or around 0 dB SPL or below, as this is where a true anechoic environment is required. Rather, most loudspeaker measurement systems are required to characterise real-world sound pressure levels of performance, ideally over the 5 Hz to 50 kHz range. Modern measurement systems can easily see into any noise and isolate it from the true acoustic output, so does it really make sense to focus so strongly on this aspect?

First introduced in 2013, the TTC [5, 4], forming part of the AES X-223 project, [1] is a simplified concept in loudspeaker measurement. A TTC provides a unified loudspeaker measurement environment that is small, stable, repeatable, and convenient; as such, it can be used throughout the entire supply chain, from design through manufacture and quality control and onward to the final customer.

Rigidly defined measurement geometry together with interchangeable sub-baffles ensures rapid, accurate, and repeatable measurements, while the TTC's relatively small physical size means that it is readily transported if required, and easily set up by one person.

A TTC acts like a mini anechoic chamber at the highest frequencies but provides calibrated results of the loudspeaker in the chamber at low frequencies. It takes the form of a tri-rectangular tetrahedron which will fit neatly into the corner of a room. A microphone is fitted towards the back corner inside, and the speaker to be tested is placed so that it fires into the chamber (at the microphone)—not out of it like a normal loudspeaker enclosure. See figure 34.2.

The loudspeaker is connected to an amplifier that takes its input from the analysis equipment (hardware or software) that you are using. The microphone is also connected to the same equipment, allowing the test programme to check if what came out is what went in. As the arrangement always allows the test loudspeakers to be fitted in the same position with respect to the microphone, the results are both accurate and very repeatable.

The basis upon which the TTCs have been designed is to minimise errors in loudspeaker driver measurements, whether set-up inaccuracies or internal reflections. The design is based upon a tetrahedral structure with acoustic absorption at the walls/floor of the enclosure to eliminate any remaining high frequency issues.

The tetrahedral shape is more familiar as a tetrahedron constructed by four identical equilateral triangles. However, our tetrahedral shape is based upon four right-angled triangles that fit easily into a corner.

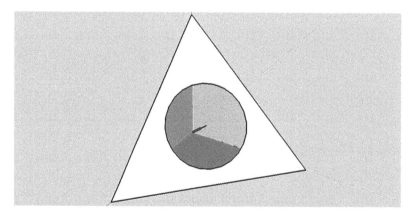

Figure 34.2: Simple Tetrahedral Test Chamber.

34.5 Equipment Used and Test Methodology

The TTC chosen for this report will be one of the medium versions (the TTC750 plus a final comparison with the TTC350). The TTC750 we used has an approximately triangular footprint of 760 × 760 mm and is 800 mm high (with feet). This version is suitable for loudspeakers up to 200 mm diaphragm diameter. For this report we will present measurements on a SEAS H1207, with a stiff, light, aluminium cone and a low-loss rubber surround.

- A theoretical boundary element model was built to test the underlying theory.
- Theoretical plots of the sound pressure level versus frequency are to be produced.
- Theoretical plots of the sound pressure field at various frequencies are to be produced.
- Measurements of a SEAS H1207 driver on an IEC baffle as a ground plane.
- Measurements of a SEAS H1207 driver are conducted using a medium TTC.
- The acoustic measurements are to be made using HOLMImpulse software, the internal and external microphone being an NTi M2010 0.5-inch type.
- Draw appropriate conclusions from the theory and measurements.
- Further measurements of another TTC350 fitted with a Behringer C2 microphone.

34.6 Theoretical Simulations on a Tetrahedral Test Chamber

A simplified BEM model of a TTC350 is shown as figure 34.3.

The results from a boundary element model (BEM) were made using ABEC 3 [2].

A section through this, demonstrating the sound pressure level distribution by colour, is shown as figure 34.4. This clearly shows that below 500 Hz the system works as a pressure environment. Above this frequency, some modes are effectively controlled by the acoustic absorption.

The simulated sound pressure level versus frequency of a chamber without damping at the microphone position is shown as figure 34.5.

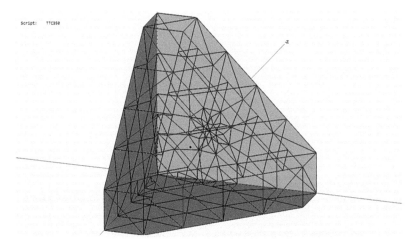

Figure 34.3: TTC350 ABEC Model.

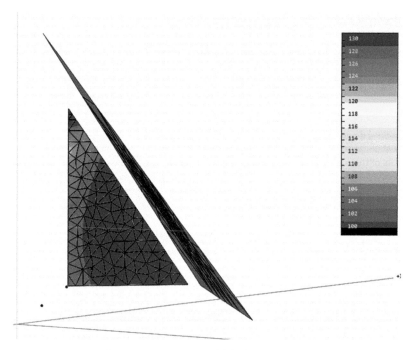

Figure 34.4: TTC350 Field Plot at 300 Hz.

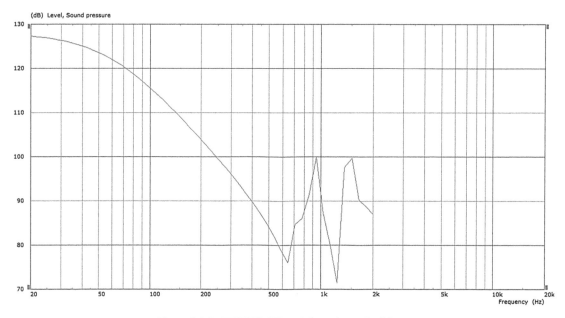

Figure 34.5: TTC350 SPL at Microphone Position.

34.7 Conventional Measurements

These measurements were made on a SEAS H1207 kindly supplied by SEAS of Norway. A full-size IEC baffle was laid out as a ground plane with a half-inch microphone (not shown) which was vertically suspended directly above the loudspeaker at approximately 1 m. A photo of a large IEC baffle as a ground plane is shown as figure 34.6.

An unwindowed IEC baffle/ground plane measurement is shown as figure 34.7.

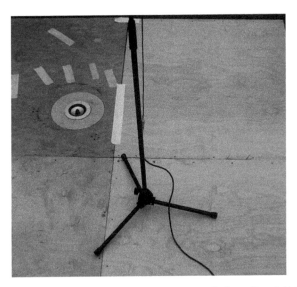

Figure 34.6: IEC Baffle Configured as a Ground Plane (Partial View).

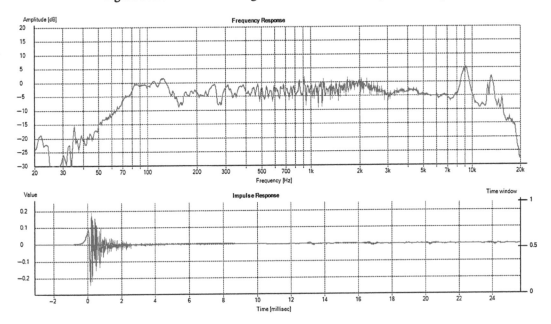

Figure 34.7: Unwindowed Measurement from IEC Baffle/Ground Plane.

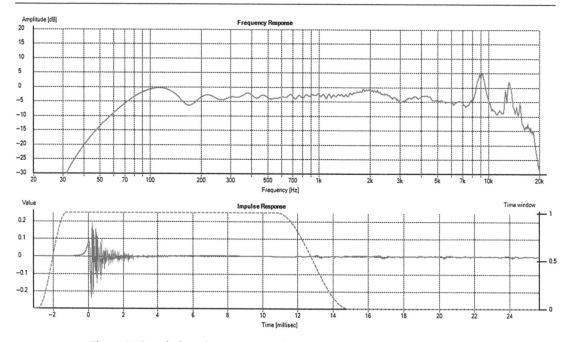

Figure 34.8: Windowed Measurement from IEC Baffle/Ground Plane at 12 ms.

As can clearly be seen, there are significant reflections; however, these mainly start after 10–12 ms, the major impulse reflection starting at around 13 ms, the low frequencies being unreliable. A 12 ms window was applied to this measurement and is shown as figure 34.8.

This clearly removes the majority of the reflection artefacts, especially above 500 Hz, where this result can justifiably be considered as anechoic. In theory a 12 ms window should be reasonably accurate below 160 Hz and is often considered safe down to 80 Hz! We can also clearly see the influence of diffraction at edges of the IEC baffle, proving that even a baffle of this size is far from perfect.[2]

34.8 Tetrahedral Measurements

The unwindowed internal measurement is shown as figure 34.9.

As can clearly be seen, there is a considerable rise in energy at low frequencies; this can also be seen in the lower time trace. The response from 2 kHz upward looks fairly good. As most loudspeaker drivers are predominantly pistonic at lower frequencies, the front of a loudspeaker diaphragm will be 180° out of phase with the rear of the diaphragm where this applies. We can use this fact by measuring the diaphragm output in the very near field (5–10 mm from diaphragm).

This is shown as the unwindowed external measurement, figure 34.10.

As can clearly be seen, this response follows very closely a simulated response, at least up to approximately 1700 Hz, which corresponds to the theoretical pistonic range for a 10 mm cone depth with a diameter of around 100 mm.

We will overlay the external measurement with that of the windowed data; this is shown as figure 34.11.

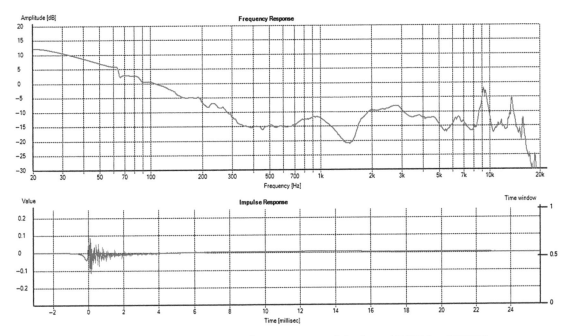

Figure 34.9: Unwindowed Internal Measurement of the Same H1207 in a TTC750.

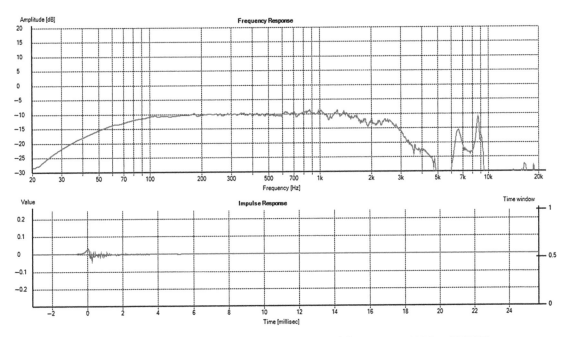

Figure 34.10: Unwindowed External Measurement of the Same H1207 in a TTC750.

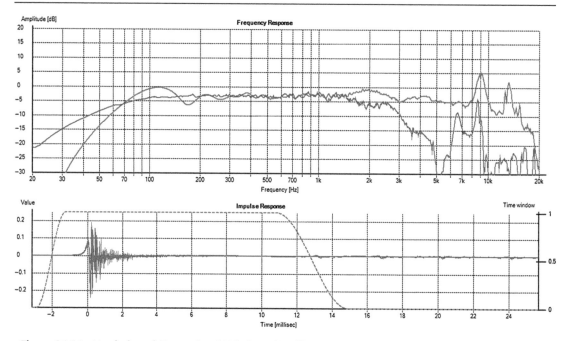

Figure 34.11: Unwindowed External and Windowed Baffle Measurement of the Same H1207 in a TTC750.

We can clearly see the closeness between these from 200 Hz to 1500 Hz. It also shows the low frequency errors from windowing, diffraction from the IEC baffle edges, and cancellation. From the work of Richard Small and Don Keele, we can use the low-frequency response internally and in the near field to predict the far field response of a loudspeaker in a given enclosure accurately. So we can accurately predict a low-frequency correction curve.

Firstly we overlay the internal and external curves, as shown as figure 34.12.

Then we will simply take the difference between these two curves.[3] In practice this can simply be achieved in a spreadsheet by subtracting the dB levels of these two curves at each frequency in turn. Note: Phase is *not* required.

This rough initial equalisation curve is shown as figure 34.13.

The resulting correction curve is showing a considerable 45 dB rise at low frequencies, but it is inaccurate at high frequencies. So we set the levels above, say, 1700 Hz to be equal.[4]

The final correction curve is shown as figure 34.14.

This then becomes the final equalisation which is applied to the internal SPL curve. The internal measurement and final correction curve are shown in figure 34.15.

We then get the result shown in figure 34.16.

The low-frequency rise has been completely eliminated. This is confirmed by the lower time trace which now tails off very rapidly, indicating at least theoretically that we can use this technique to measure accurately to very low frequencies.

The windowed IEC baffle versus TTC750 measurement is shown as figure 34.17.

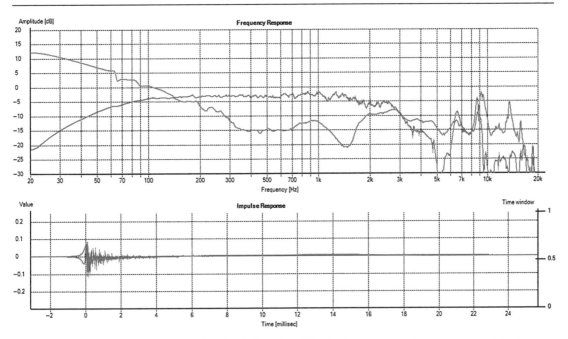

Figure 34.12: Unwindowed External and Windowed IEC Baffle/Ground Plane Measurement of the Same H1207 in a TTC750.

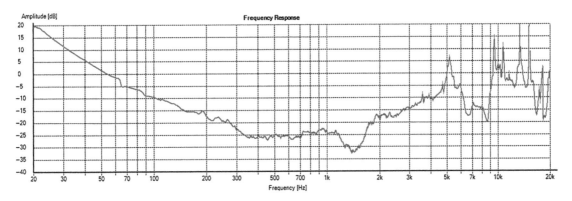

Figure 34.13: Rough Equalisation Curve for a TTC750.

Clearly the low-frequency response is accurate. However, the high-frequency performance is also quite reasonably reliable within 5 dB of the windowed response, even without further correction, which could be applied if higher accuracy was desired.

Next we can see an overlay of windowed IEC baffle versus TTC350 and TTC750 measurements, in figure 34.18, where IEC baffle = brown, TTC350 = green, and TTC750 = blue, with normalised data using the same H1207 driver. These measurements were made at intervals of years and with different drivers, so the high frequencies are not

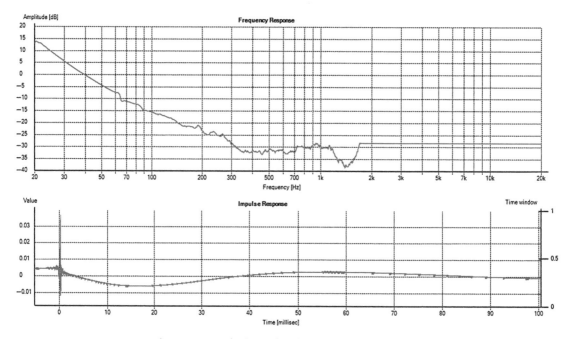

Figure 34.14: Final Equalisation Curve for a TTC750.

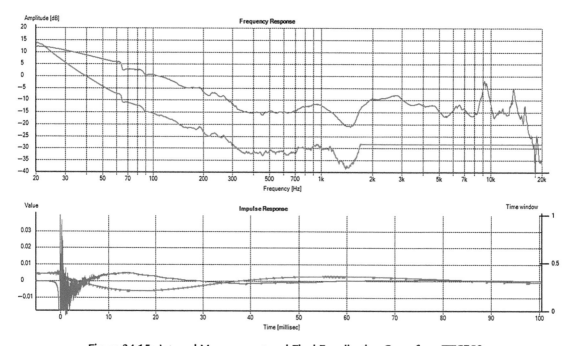

Figure 34.15: Internal Measurement and Final Equalisation Curve for a TTC750.

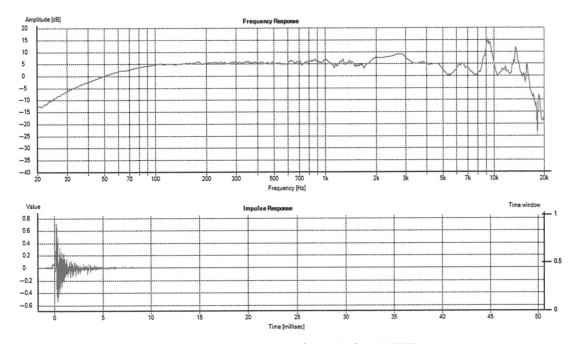

Figure 34.16: Final Result of a H1207 in a TTC750.

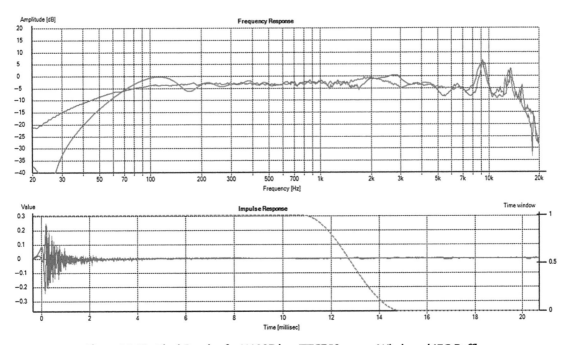

Figure 34.17: Final Result of a H1207 in a TTC750 versus Windowed IEC Baffle.

identical. However, the low-frequency measurements clearly show the different responses at 20 Hz between the different chamber sizes.

An example of a TTC750 tetrahedral test chamber is shown as figure 34.19.

An example of a TTC350 tetrahedral test chamber is shown as figure 34.20.

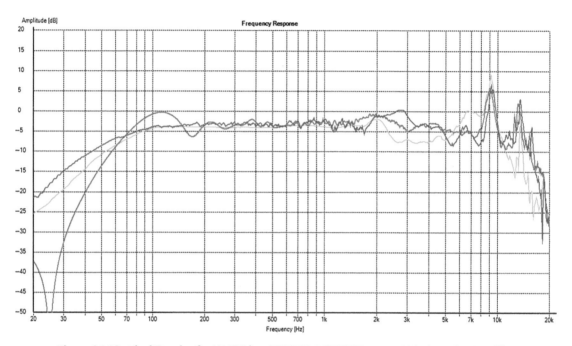

Figure 34.18: Final Result of a H1207 in a TTC350 & TTC750 versus Windowed IEC Baffle.

Figure 34.19: Tetrahedral Test Chamber—TTC750.

Figure 34.20: Tetrahedral Test Chamber—TTC350.

34.9 Measurement Equipment and Types of Loudspeakers

Since their introduction, the TTCs have been used for measurements of the following:

- Microspeakers in China.
- Sounders in Germany.
- Tweeters in the U.K.
- Mid-bass loudspeaker drivers in China, the U.K. and the U.S.A.
- Subwoofer bass drivers in France.
- Complete loudspeaker systems in the U.K. and India.
- Mobile phones in the U.K. and the U.S.A.

Using equipment from the following:

- Audio Precision.
- Audiomatica CLIO, CLIO QC, and CLIO Pocket.
- Bruel & Kjaer.
- Etani.
- Klippel, R&D, and QC systems.
- Loudsoft FineQC.
- Listen.
- Nti-audio.

34.10 Comparison and Discussion

Comparing the final correct response curve with the IEC curve shows a remarkably good correlation at high frequencies with known calibrated measurements at low frequencies. All the critical features of the loudspeakers' frequency response are captured by the TTC.

The rigidly defined measurement geometry together with the interchangeable sub-baffles of the chambers ensures rapid, accurate, repeatable measurements time and time again. The chambers can be easily calibrated and either a generic response correction used or a specific correction produced for when the highest accuracy is required.

Due to the comparatively small size of the various TTCs, it is possible to use them in situations where a full anechoic chamber and/or IEC baffle could never be used, such as the designer's bench, QA/QC lab, end of the production line, and goods received test station. When used throughout the supplier chain, results are directly comparable, eliminating major communication problems.

34.11 Conclusion

The measurements from the TTCs correlate very well with those measured by an IEC baffle; however, they are of much higher resolution and do not show the diffraction problems of IEC baffle measurements.

From the both a theoretical point of view and from actual physical measurements, the TTCs have proved capable of giving accurate, repeatable measurements within a relatively small size.

They can be easily calibrated and either a generic response correction used or a specific correction produced for when the highest accuracy is required very simply using the steps in this report. Such a chamber would dramatically simplify setting measurement tolerances on loudspeakers by eliminating practically all of the modal effects of our current loudspeaker test enclosures.

This report has proven the capability of the TTC to make substantial improvement in our acoustic loudspeaker measurements. We should embrace this technology throughout our organisation.

Use of several of the TTCs throughout the entire supply chain would give our companies unprecedented accuracy and consistency of acoustic measurements. This will lead to lower costs and higher standard products. By eliminating the current situation of differing acoustic measurement set-ups at different stages and replacing them with a single known set-up capable of the highest standard, we can make accurate comparative measurements throughout the supply chain.

By using these at all stages of production we will eliminate the unknown variables that are purely due to the different measurement techniques currently in use.

TTCs are available from Hill Acoustics http://hillacoustics.com/.

Notes

1. At 20 Hz, being approximately 17 m.
2. The detail of how this was done is presented in Appendix F (HOLMImpulse Tutorial) on page 251.
3. The exact details differ between various measurement systems.
4. This may also be used as the 0 dB reference level. The exact method will depend upon the measurement system.

References

[1] www.aes.org/standards/meetings/init-projects/aes-x223-init.cfm (visited on 06/02/2018).
[2] *ABEC*. www.randteam.de/ABEC3/Index.html (visited on 02/02/2018).
[3] Leo L. Beranek and Harvey P. Sleeper Jr. "The design and construction of anechoic sound chambers". In: *J. Acoust. Soc. Am.* 18.1 (1946), pp. 140–150.
[4] Geoff Hill. "Comparative Results between Loudspeaker Measurements Using a Tetrahedral Enclosure and Other Methods". In: *Audio Engineering Society Convention 136*. Apr. 2014. www.aes.org/e-lib/browse.cfm?elib=17152.
[5] Geoff Hill. "Consistently Stable Loudspeaker Measurements Using a Tetrahedral Enclosure". In: *Audio Engineering Society Convention 135*. Oct. 2013. www.aes.org/e-lib/browse.cfm?elib=16958.

[6] D. B. (Don) Keele Jr. "Low-Frequency Loudspeaker Assessment by Nearfield Sound-Pressure Measurement". In: *J. Audio Eng. Soc* 22.3 (1974), pp. 154–162. `www.aes.org/e-lib/browse.cfm?elib=2774`.

[7] Alan S. Phillips. *Measuring the True Acoustical Response of Loudspeakers*. Tech. rep. SAE Technical Paper, 2004.

[8] Richard H. Small. "Simplified Loudspeaker Measurements at Low Frequencies". In: *J. Audio Eng. Soc* 20.1 (1972), pp. 28–33. `www.aes.org/e-lib/browse.cfm?elib=2103`.

PART IX

Appendices

Glossary

Term	Description
α	System compliance ratio ($C_{ms}/C_{mb} \equiv C_{as}/C_{ab} \equiv V_{as}/V_{ab}$) of the driver, divided by cabinet compliances in mechanical, acoustical, or volume terms.
$\circ$	Phase (degrees), used to describe either the starting position of a sine wave or the relative phase difference between two waveforms.
$^{\circ}$C	The symbol used to define temperature in degrees Celsius.
Δ	Increment, a term used to define a normally small change in a parameter.
ϵ	Epsilon, lower case, used to define the extensional strain in Young's modulus.
$\equiv$	Identical to, used for an equation whose parts balance out for all possible values.
η_0	Passband Efficiency %. In acoustic terms this needs to take account of the acoustic load at frequencies of interest. So if a source is well within ($\ll$) a given wavelength, then the presence of one boundary will cause a corresponding increase in output or efficiency.
λ	Lambda, lower case, or wavelength, used to define the length of a waveform from one peak or dip to the next one. It's usually measured in metres (m).
μ	Micro or 1 part in 1,000,000, usually denoted by the Greek symbol μ.
μ_0	μ_0 is used to define the magnetic constant formerly and often still called the permeability of free space. This is a defined unit where $\mu = 4 \times \pi \times 10^{-7}$ or $\approx 1.2566370614\ldots \times 10^{-6}$ (N/A^2).
μ_R	Relative permeability, the ratio of a material permeability to the permeability of free space μ_0.
ν	Nu, lower case, used to denote Poisson's ratio.
π	Ratio of circumference to diameter of a circle. An irrational number that can never be calculated precisely, it's usually taken as $3.141592765359\ldots$
ρ_0	Density of air, (1.18 kg/m^3 at sea level). As the density varies greatly with height, also with atmospheric pressure (nominally 1000 mB) and temperature 20 $^{\circ}$C, it is critical as the speed of transmission of any waveform is highly dependant upon the density of the medium it transverses.
σ	Sigma, lower case, is used as a symbol for the standard deviation of a population or probability distribution.
Σ	Sigma, upper case, is used as a symbol for the summation operator.
ω	Angular or radian frequency, $2 \cdot \pi \cdot f$ (Hz)
0 dB	Logarithmic ratio of identical quantities, zero dB.
0 dB SPL	The nominal threshold of human hearing is normally taken to be the peak sensitivity region of 2–3 kHz, where 0 dB SPL equals 20 μPa.
a	Distance from source to measuring point (m).
A.C.	Alternating current (amperes), usually referring to a sinusoidal waveform.

Continued on next page

Term	Description
AlNiCo	Alnico, referring to aluminium, nickel, and chromium: the base elements of a type of permanent magnet.
Anisotropic	A material whose properties change according to the direction in which it is stressed or measured.
B	Magnetic flux density, usually measured in a gap but can be measured within a material, in which case it can show saturation. Measured in Tesla (T).
BEM	Boundary element method, one of the mathematical and computational techniques used to break down a complex problem so that it can be solved by summing many individual solutions.
$BH_{product}$	Magnetic energy, equals the magnetic flux density×magnet field strength at any point on the demagnetisation curve.
BH_{max}	Magnetic energy, equals the magnetic flux density × magnetic field strength at the maximum point of the demagnetisation curve.
Bl	Product of flux density and length of wire in the same magnetic field, measured in (Tm).
$Bl(x)$	The variation of Bl with x, which is the displacement, usually in mm.
Br	Residual induction (or flux density), the magnetic induction remaining in a magnet after saturation in a closed circuit. Measured in tesla (T).
BTL	Bridge tied load, used to denote an output from an amplifier comprising two outputs normally inverted to each other, the load being connected to both amplifier outputs and not to a ground connection.
c	Speed of sound in air, (343 m/s), normally taken in dry air at a temperature of 20 °C.
C_{ab}	Acoustic compliance of the air in cabinet (m^5/N).
C_{as}	Acoustic compliance of a loudspeaker drive unit (m^5/N).
C_{at}	Acoustic compliance of loudspeaker system (m^5/N).
C_{ms}	Mechanical compliance of a loudspeaker driver's moving parts, $\frac{C_{as}}{S_d^2}$.
$C_{ms}(x)$	Mechanical compliance of a loudspeaker driver's moving parts with displacement (m/N).
CAD	Computer aided design or drawing, whereby drawings are made digitally on a computer.
CAM	Computer aided manufacturing, whereby component parts are made with computer assistance.
D.C.	Direct current, a uni-directional flow of current in a circuit, measured in amperes (A).
D_v	Diameter of vent or port (m).
dB	Decibel or $1/10^{th}$ of a bel. A logarithmic ratio of two quantities allowing very large and very small quantities to be visually displayed. It also matches quite closely the dynamic capabilities of the ear; it is normally taken as the logarithm to base 10.
dwg	A common AutoCAD computer aided drawing format.
dxf	Interchangeable drawing file format.
f	Frequency, the number of occurrences of a repeating event in a given unit of time (normally one second). Measured in hertz (Hz).
f_c	Closed box resonant frequency (Hz).
f_s	Driver resonant frequency in free air (Hz).
FEA	Finite element analysis, one of the mathematical and computational techniques used to break down a complex problem so that it can be solved by summing many individual solutions.
$G(s)$	Function G of s, the relationship of a system's output to its input in the complex s domain; thereby taking simultaneous account of frequency and phase.
Hc	Coercive force is the ability of a magnetic material to withstand an external magnetic field without de-magnetisation occurring.

Continued on next page

Term	Description
I	Current, (I), the international symbol given to current, which is measured in amperes (A).
$I(x)$	Current versus displacement (mm).
Impulse	A theoretical Impulse or DIRAC pulse is a uni-directional spike that contains all frequencies simultaneously, theoretically of infinite amplitude, infinitesimal rise and fall times, and of negligible duration. In practice we fall short of this ideal, and practical impulses are of limited but useful bandwidth.
Impulse Response	The term used to describe the reaction of any system in response to an impulse.
Isotropic	A material whose properties are uniform with respect to direction.
j	$\sqrt{-1}$, the imaginary number used to represent the phase variance needed to manage complex numbers with amplitude and phase. (Mathematicians tend to use i, but this is easily confused with current.)
K_{ms}	Stiffness of the mechanical suspension, the inverse of mechanical compliance: $\frac{1}{C_{ms}}$.
$K_{ms}(x)$	Stiffness of the mechanical suspension versus displacement x, the inverse of mechanical compliance with respect to x: $\frac{1}{C_{ms}(x)}$.
L	The international symbol used for inductance. Named after Emil Lenz, it is measured in henrys (H) (after Joseph Henry).
l	Length of wire in magnetic air gap (m).
L_v	Length of vent or port (m).
L_e	Voice coil inductance (H).
$L_e(x)$	Voice coil inductance versus displacement.
m	Metre.
mm	Millimetre or 1 part in 1,000.
M_{as}	Acoustic mass of driver. Normally this includes a contribution from the mass of the air loading the diaphragm.
M_{at}	The total acoustic mass of system.
M_{av}	Equivalent acoustic mass of vent or port.
M_{ms}	Mechanical mass of driver, $\frac{M_{as}}{S_d^2}$.
MIL-STD	Military Standard.
Newton	S.I. unit of force (N), the force required to accelerate a mass of one kilogram at the rate of a metre per second squared in the direction of the applied force.
NdFeB	Magnet made from an alloy of neodymium, iron and boron.
ohm	S.I. Symbol of electrical resistance (Ω), named after George Ohm.
Pa	Pascal, SI unit of pressure.
Q	Ratio of resistance to reactance for a parallel or serial circuit.
Q_a	Q factor defining absorption losses for cabinet at f_b or f_c.
Q_b	Q factor defining losses for cabinet at f_b or f_c.
Q_{es}	Electrical driver Q at f_s.
Q_l	Q factor defining losses for cabinet leakage at f_b or f_c.
Q_L	Quality factor.
Q_{ms}	Mechanical driver Q at f_s.
Q_{tc}	Q factor defining total driver and cabinet losses at f_c.
Q_{ts}	Q factor defining total driver losses at f_s.

Continued on next page

Term	Description
R_{ab}	Acoustic resistance of box absorption losses.
R_{al}	Acoustic resistance of box leakage losses.
R_{as}	Acoustic resistance of driver suspension losses.
R_{at}	Acoustic resistance total, $R_{as} + B^2 \cdot l^2 / [(R_g + R_e) \cdot S_d^2]$.
R_b	$1/(Q_b \cdot \omega_c \cdot C_{ab})$.
R_e	DC or zero frequency voice coil resistance (Ω).
$R_e(T_v)$	Voice coil resistance temperature variation ($\Omega/°C$).
R_{ms}	Mechanical resistance of driver suspension losses, $(R_{ms} \cdot S_d^2)$.
R_2	Motor unit semi resistance (Ω).
s	Complex variable or Laplace transform, $(\sigma + j\omega)$.
S_d	Effective surface area of the driver (m^2).
SmCo	Samarium Cobolt.
SPL	Sound Pressure Level (dB SPL) normally re 20 μPa.
$SrFe_2O_2$	Hard ferrites or ceramics.
T_v	Temperature ($°C$).
T	Tesla, S.I. unit for flux density (T).
USB	Universal serial bus.
V	Voltage (V)
V_{as}	Volume of air with acoustic compliance, $(\rho_o \cdot c^2 \cdot C_{as})$.
V_b	Volume of cabinet (m^3). This is normally stated in litres.
X	3D coordinate X direction.
Y	3D coordinate Y direction.
Z	3D coordinate Z direction.
Z	Impedance (Ω).
Z_{max}	Maximum impedance (Ω).
Z_{mean}	Calculated mean impedance (Ω).
Z_{min}	Minimum impedance (Ω).

Resources

B.1   Books

- *Applied Acoustics, second edition* (1939). Olson, Harry and Massa, Frank. (P. Blakiston's Sons & Co. Inc.)
- *Audio Transducers* (2002). Geddes, Earl and Lee, Lidia. (GedLee Associates LLC), ISBN 978-0-9722085-0-5
- *Electronic Design, fourth edition* (2002). Roden, Martin S., Carpenter, Gordon L., and Wiesserman, William R. (Discovery Press), ISBN 0-9646969-1-6
- *Engineering Mathematics, seventh edition* (2013). Stroud, K.A. and Booth, Dexter J. (Palgrave Macmillan), ISBN 978-1-137-03120-4
- *Advanced Engineering Mathematics, fifth edition* (2011). Stroud, K.A. and Booth, Dexter J. (Palgrave Macmillan), ISBN 978-0-230-27548-5
- *High Performance Loudspeakers, sixth edition* (2005). Colloms, Martin. (Wiley), ISBN 978-0-470-09430-3
- *Loudspeakers* (1963). Jordan, E.J. (Focal Press)
- *Sound Reproduction, third editon* (1953). Briggs, G.A. (Wharfedale Wireless Works)
- *Sound Reproduction: Loudspeakers and Rooms* (2008). Toole, Floyd E. (Taylor & Francis) ISBN 978-0-240-52009-4
- The Audio Engineering Society www.aes.org has published many important papers referred to in this book including the following anthologies:
  Loudspeakers Vol.1, 2nd edition (1980)
  Loudspeakers Vol.2, 1st edition (1984)
  Loudspeakers Vol.3: Systems and Crossover Networks, 1st edition (1996) ISBN 0-937803-28-6
  Loudspeakers Vol.4: Transducers, Measurement and Evaluation, 1st edition (1996) ISBN 0-937803-29-4

B.2   Websites

All of these websites were working as of March 2018. If in the future they are not, try looking in a web archive like: Wayback Machine: http://archive.org/web/web.php

- ABEC, AxiDriver & VACS: www.randteam.de
- Acoustical Society of America: www.acousticalsociety.org
- ANSYS®: www.ansys.com
- ARTA: www.artalabs.hr
- Audio Engineering Society: www.aes.org
- Autodesk® FUSION 360™: www.autodesk.com/products/fusion-360/overview
- CAE linux: www.caelinux.com/CMS/
- Calculix: www.calculix.de
- COMSOL Multiphysics®: www.comsol.com
- DraftSight®: www.3ds.com/products/draftsight/overview/
- Excel® Viewer: www.microsoft.com/en-us/download/details.aspx?id=10
- FEMM: www.femm.info
- FreeCAD: http://free-cad.sourceforge.net/

- FreeMat: http://freemat.sourceforge.net/#home
- Hill's site for SpeakerPro, Theoretical Bass Design & Theoretical Bl(x): www.geoff-hill.com/
- HOLMImpulse: www.holmacoustics.com/holmimpulse.php
- Klippel: www.klippel.de
- www.linkwitzlab.com/
- Linkwitz Transform: http://sound.whsites.net/linkwitz-transform.htm#how-works
- LISA: www.lisa-fet.com
- MecWay: http://mecway.com/
- Micro-Cap: www.spectrum-soft.com/index.shtm
- Octave: www.gnu.org/software/octave/
- OnShape: www.onshape.com/
- OpenOffice™: www.openoffice.org
- OpenFOAM: www.openfoam.org
- OpenProj: http://sourceforge.net/projects/openproj/
- Pafec and PafLS: www.pafecfe.com
- Salome: www.salome-platform.org
- Scilab: www.scilab.org/
- Sketchup: www.sketchup.com/products/sketchup-make
- Speak: www.gedlee.com/SPEAK/Speak.aspx
- Strand7®: www.strand7.com/html/demosoftware.htm
- Visual FEA: www.visualfea.com/
- WinISD: www.linearteam.org

ABEC and VACS Tutorial

This tutorial on ABEC (Acoustic Boundary Element Calculator) builds up a mechanical and acoustic model of our subwoofer; the acoustical output is then graphed using VACS (Visualising Acoustics). Both are available at www.randteam.de.

ABEC works primarily in a 3D mode and X, Y, or Z can be selected as planes of symmetry. It is controlled by simple text files. VACS can be used to read and manipulate pretty well any standard acoustic measurement file, as well as providing the graphing function for ABEC. ABEC can also read mesh files compiled with the assistance of 3D CAD (geometry) and mesh files by gmesh or data files from external sources.

C.1 Initial Setup

However, before you start we would *very strongly* recommend using a good text editor to display the working code. Figure C.1 shows the nodes text in Windows Notepad on the left and in Notepad++ on the right. Clearly, the code displayed in Notepad++ is much clearer and easier to read and understand.

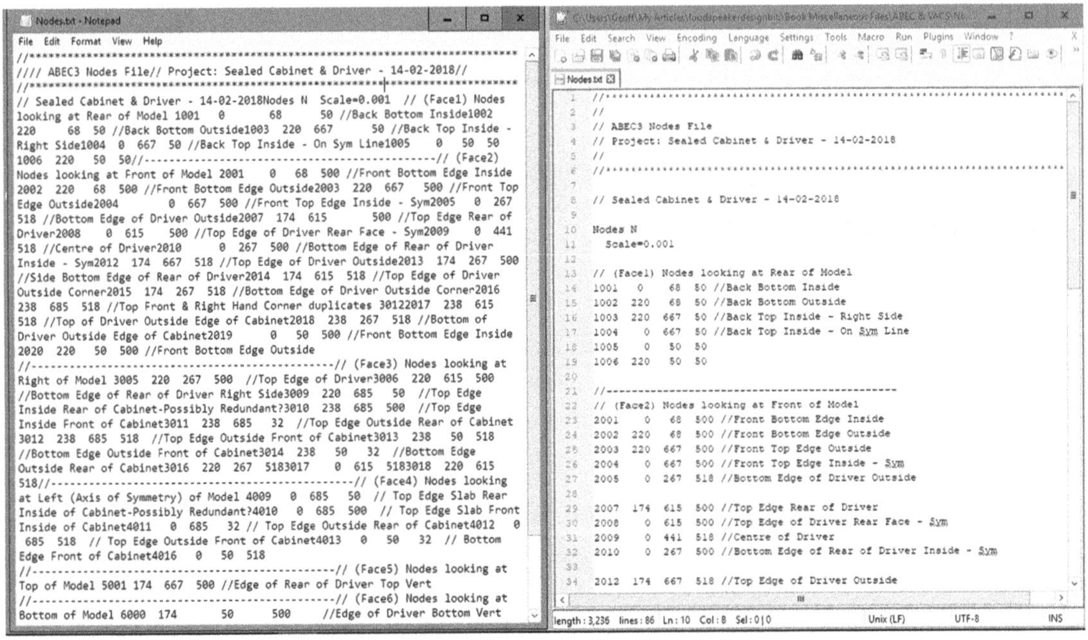

Figure C.1: Nodes in Notepad and Notepad++.

When ABEC is opened up, one is presented with the main screen, as shown in figure C.2.

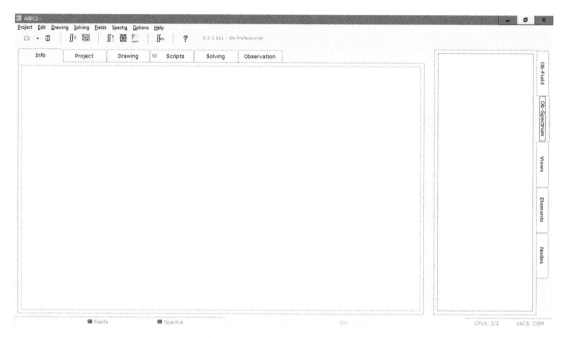

Figure C.2: ABEC 3 Main Screen.

Next we will enable Notepad++ as the default editor. Click on Options->Preferences->Editor and enter the location of your chosen text editor. This is shown in figure C.3.

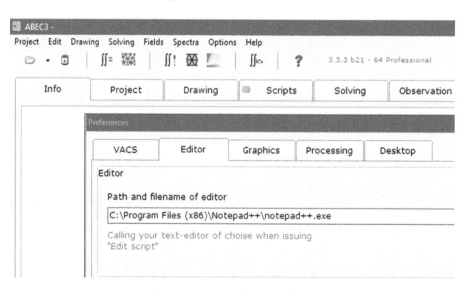

Figure C.3: Text Editor Location.

C.2   Modelling Scripts

We will now look at constructing a model from scratch. Confusingly, ABEC does not have a way of directly creating a script file, so you need to either create one using the text editor or open an existing sample file. The next step is to attach the individual script files to a project file. This can be done either in a text editor or in the ABEC main screen under the Project tab as shown in figure C.4.

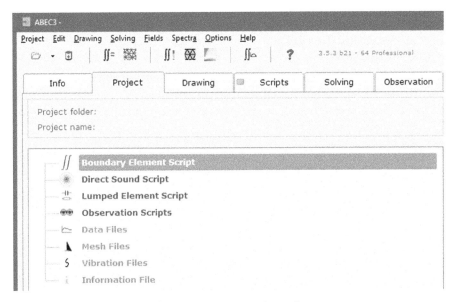

Figure C.4:   ABEC Project Tab.

The key scripts normally recommended are: Boundary Element Script, Direct Sound Script, Lumped Element Script, and Observation Scripts.

We prefer to break this down a little more as follows:

1.   Nodes.
2.   Solver.
3.   LEScript.
4.   Observation Spectrum.
5.   Observation Field.
6.   Formula.[1]

Initially, we need to make our first ABEC project file, which is easy using the normal 'Save As' dialogue. The ABEC project file is just a text file. It uses a .abec file extension but can be examined by a normal text editor. When a project is first opened up it is completely blank, so the second task is to attach the relevant scripts to the project file. There are two choices here:

1.   If you have the project file opened in a text editor, simply add the name of the file after the equals sign '='.
2.   If you are using the ABEC main screen, click the Project tab, highlight the name of the type you want to attach (e.g. 'Solver'), right click on it and navigate using the browser to the file you want to attach.

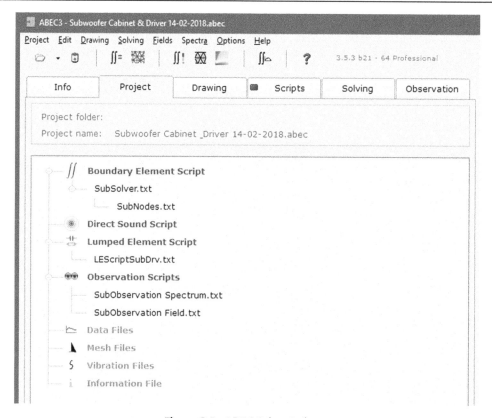

Figure C.5: ABEC Select Scripts.

If one of these files needs to call other files, the best way to include these is by using the statement 'File=' followed by the name/path of the file. Note that ABEC expects all files to be in the same (current) directory. To override this behaviour, feed it a full name and path. These are shown in figure C.5.

C.3 Begin Modelling

Now we know how a project is structured in ABEC, we can begin modelling. The first thing to do is to prepare a 3D sketch of the object shape or environment that we wish to model. We then need to mark up the main nodes on this.

The second is to produce a list of the key dimensions of the object we wish to model: key dimensions such as height, width, and depth, together with thickness, driver and vent dimensions, parameters, and so on.

The third is to produce a series of six sketches of our object from the six main directions of a cube, again with the nodes marked on them. (Theoretically you only need a single sketch but this can quickly get very complicated.)

We can then break these down into a series of nodes with their appropriate X, Y, and Z dimensions. ABEC uses these nodes later for constructing the geometry. It is essential that these nodes are unique so let us decide that the front nodes begin with 1, the back ones with 2, and so on.

In ABEC, put any comment after a pair of forward slashes //. All the nodes in ABEC are unique and must use numbers. However, there is nothing to prevent you assigning several nodes to the same X, Y, and Z coordinates. It can be difficult to assign boundaries or planes to these later as the node numbering will also overlap, making it difficult to distinguish them from each other. After a while you end up with something like this:

```
Nodes N
Scale=0.001 // converting from mm to m
```

As shown in table C.1.

Table C.1: Initial Nodes.

Node No.	X	Y	Z
101	0	0	0
102	0	0	200
103	0	200	200
104	0	200	0

```
//**********************************************************************
// ABEC3 Nodes File
//**********************************************************************

101    0      0        0
102    0      0      200
103    0    200      200
100    0    200        0
//**********************************************************************
```

Here we have four coordinates, and in the Solver section we can stitch them together as a structural element #1000 created from nodes #100, #101, #102, and #103. This should produce a 'square' of 200 mm × 200 mm, as shown as figure C.6.

If we look at the drawing view, we should clearly see it. If we extend the 'Normals' look at node #102, you can just see the 'line' representing the normal direction, so we can see which direction it is working in. If it is in the wrong direction for us, then we need to change the order of the nodes creating element #1000: 103 102 101 100.

Sometimes, however, instead of a quadrilateral shape you end up with two crossed triangles over the shape. In this case, the order of the individual nodes is confused and does not rotate correctly. To correct this, the order of the nodes themselves needs to be changed.

ABEC uses a 'right hand' rule to work out the direction nodes should be joined up in. Take your right hand, hold it with the thumb extended upward (this corresponds to the direction of the normal or the direction of the plane being upward), and curl the fingers back toward the palm in an anticlockwise direction. The individual nodes need to follow this direction to match up with an upward facing plane.

C.4   Modelling Our Subwoofer

This is fine for simple models but soon gets out of hand with more complicated examples. Fortunately, ABEC allows the substitution of exact numbers by variable names and this is what we will use a formula script for.

In the case of our subwoofer, we know the following: width = 450 mm, depth = 450 mm, height = 450 mm, thickness = 25 mm, driver area = 950 cm²; therefore, as area = $\pi \cdot r^2$, so $r^2 = 950$ cm² and $r = \sqrt{\frac{0.31}{3.141259}} = 348$ mm.

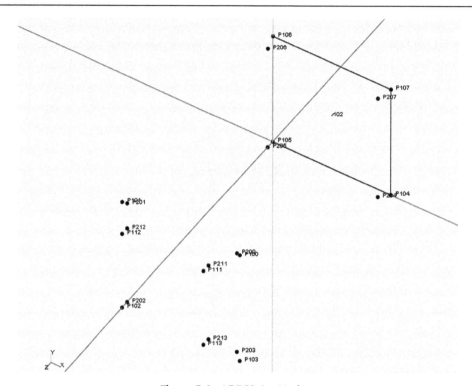

Figure C.6: ABEC3 1st Nodes.

C.5 Formula

Let us set up a formula to describe this loudspeaker. We need to start off with a name for the formula, and when we want to call these names later, we open curly braces to contain the descriptors.

```
//***********************************************************************
// ABEC3 Nodes File
// Project: Sub Cabinet & Driver - 07-02-2018
//***********************************************************************
Formula //"Cabinet"
T=25                    //Cabinet Thickness
H=450/2                 //Cabinet Height
D=450/5                 //Cabinet Depth
W=450/2                 //Cabinet Width/2
Dr=348/2                //Driver Diameter
Dh=H/2                  //Driver Height
B=-225                  //Base Height
Wd=100                  //Distance from Back wall
G=2.5                   //Meshing Gap
IG=2.5                  //Internal Mesh Gap
EdgeL=500

dD=Dr*2/1000            // Diameter of cone (337mm + 11mm suspension)
```

```
tD1=120/1000//mm          // Inner depth of cone to base of dust cap
dD1=100/1000//mm          // Diameter of dust cap
dDr=(dD1/2)*1000
hD1=16/1000//mm           // Height of dust cap
hD2=95/1000//mm           // Height of total cone on the outside
dVC=75/1000//mm           // Diameter of voice coil
hVC=32/1000//mm           // Height of voice coil
dM=150/1000//mm           // Diameter of magnet
hM=20/1000//mm            // Height of magnet
t1=-18.0/1000//m          //Cone Thickness
```

We now have 22 variables that describe various parts of the cabinet or drive unit. These can be passed to other scripts that comprise the model, saving us from entering them individually when they are required.

C.6   Nodes

Next we have the 'nodes' file. This contains the structure that Solver works on for the model. However, it does not include the output results in any way; these are solved separately according to the particular requirements that need to be modelled.

We then continue building our nodes but using the variables contained in 'SubWooferCabinetFormula.txt' instead of fixed numbers. This makes it much quicker and easier to make changes later.

Also, do not forget that we have a fully symmetrical cube with a driver on one face. We can therefore use symmetry in two planes which will reduce the size to 1/4 and with 1/16 of the number of calculations required compared to a full model.

```
//********************************************************
// ABEC3 Nodes Text File
// Project: Sub Cabinet & Driver
//********************************************************
File=SubWooferCabinetFormula.txt //this loads the text file into the
                                 Nodes File
Nodes N
Scale=0.001// This allows us to use (mm) instead of (m)
```

Anyway, let's get on to the nodes themselves.

```
//**************************************************************************
// ABEC3 Nodes File
// Project: Sub Cabinet & Driver - 15-02-2018
//**************************************************************************
// Sub Cabinet & Driver - 15-02-2018
Formula "SubWooferCabinet"
{
W=450/2       // Width of Cabinet divided by 2
D=450/2       // Depth of Cabinet divided by 2
H=450/2       // Height of Cabinet divided by 2
T=25.0        // Thickness of Cabinet Walls
Di=348        // Diameter of Drive Unit
R=Di/2.       // Radius of Drive Unit
M=10          // Meshing Gap
}
```

```
Nodes N
Scale=0.001
//We then continue building our nodes but using the above formula instead of
//fixed numbers, this then makes it much quicker and easier to make changes
//later...
//Also lets not forget we have a fully symmetrical cube with a driver on one
//face we can therefore use symmetry in two planes which will reduce the size
//to 1/4 with 1/16 of the number of calculations required compared to a full
//model.

// Top of Model X,Y & Z
100  {W}          {H}          {D}          //External Top Back corner
101  0            {H}          {D}          //External Top Back corner
102  0            {H}          0            //External Drive Unit Centre

111  {W-(W-R)}    {H}          {D-(D-R)}    //External Top Corner of Driver
112  0            {H}          {D-(D-R)}    //External Top Edge of Driver on Y Axis
113  {W-(W-R)}    {H}          0            //External Top Edge of Driver on X Axis

103  {W}          {H}          0            //External Top Front corner
104  {W}          0            0            //External Top Side corner
105  0            0            0            //External Rear Centre Line
106  0            0            {D}          //External Top Back Axis corner
107  {W}          0            {D}          //External Top Back Axis corner

200  {W-T}        {H-T}        {D-T}        //Internal Top Back corner
201  0            {H-T}        {D-T}        //Internal Top Back corner
202  0            {H-T}        0            //Internal Drive Unit Centre

211  {W-(W-R)}    {H-T}        {D-(D-R)}    //Internal Top Corner of Driver
212  0            {H-T}        {D-(D-R)}    //Internal Top Edge of Driver on Y Axis
213  {W-(W-R)}    {H-T}        0            //Internal Top Edge of Driver on X Axis
214  {W-(W-R)}    {T}          0

203  {W-T}        {H-T}        0            //Internal Top Front corner
204  {W-T}        {T}          0            //Internal Top Side corner
205  0            {T}          0            //Internal Front Centreline
206  0            {T}          {D-T}        //Internal Top Back Axis corner
207  {W-T}        {T}          {D-T}        //Internal Top Back Axis corner
```

C.7 Solver

```
//********************************************************
// ABEC3 Solving File
// Project: Sub Cabinet & Driver
//********************************************************
// Sub Cabinet & Driver
File="SubNodes.txt"
File=SubWooferCabinetFormula.txt.
```

```
Control_Solver
  f1=20;   f2=500;   NumFrequencies=10;   Abscissa=log;
//*********************************************************
```

The field calculations which calculate the whole 3D field at different frequencies will use start frequency (f1) = 20 Hz and stop frequency (f2) = 500 Hz, using ten frequencies on logarithmic spacing.

```
ImpedanceNumFrequencies=100
```

The spectrum calculations are made at specific points of the 3D field and are therefore quicker. So a denser calculation is both possible and necessary, and so we use the variable 'ImpedanceNumFrequencies'.

```
Sym=XY
```

The Sym variable defines which planes of symmetry are being used in a model. Here the model is symmetrical in the X and Y planes.

```
MeshFrequency=1500Hz.
```

This partially controls the resolution of the FEA solver. As a general rule, it should be higher than the highest frequency that you need to calculate.

```
Meshing=Delaunay // This defines the shape of the triangles
Altitude=10m;   Temperature=21C
Dim=3D // Defines the model to be 3D but it could also be axisymmetric.

Subdomain_Properties "Chamber"
EdgeLength=100mm
Subdomain=1
// Here we define subdomain 1 to be "Chamber" . Also we use "edge length" to
// ensure that the mesh triangles have a maximum length of 100mm for this
// domain.
Subdomain_Properties "Exterior"
Subdomain=2
ElType=Exterior
//IBPlane=y
//IBPlane=z


Elements "Chamber"
RefNodes="N"
SubDomain=1
```

We then enable the cabinet planes. We now need to include the driver but we will change the cone diameter from a fixed value to Di/1000 so that we can change it and the cabinet together. Back in the Solver section again...

```
Diaphragm   "Cone front"
```

And finally the driver. Notice how only 1/4 of the driver is shown as the model has been instructed to use symmetry in two axes. If we check the Symmetry box in ABEC we can see that this is so as shown in figure C.7.

We have collected all of these as table C.2 on page 198.

Table C.2: Sub Woofer Nodes.

Element	Node 1	Node 2	Node 3	Node 4 (optional)	Comments
100	100	101	102	103	External Front Panel Whole not used
100	112	111	100	101	External Front Top Panel
101	111	113	103	100	External Front Side Panel
101	102	103	104	100	External XZ Symmetry Panel not used
102	104	105	106	107	External Back Panel
103	103	104	107	100	External 'Right Hand' Panel
104	107	106	101	100	External 'Top Panel'
200	203	202	201	200	Internal Front Panel Whole not used
200	201	200	211	212	Internal Front Top Panel
201	200	203	213	211	Internal Front Side Panel
201	202	203	204	205	Internal XZ Symmetry Panel not used
202	207	206	205	204	Internal Back Panel
203	200	207	204	203	Internal 'Right Hand' Panel
204	200	201	206	207	Internal 'Top Panel'

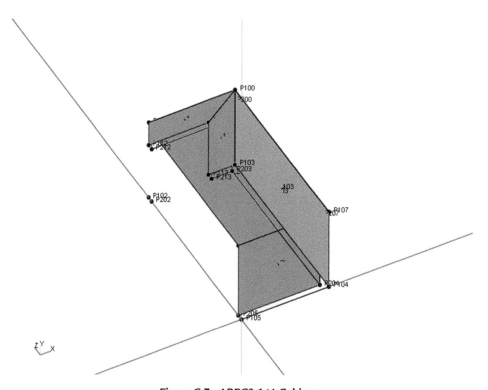

Figure C.7: ABEC3 1/4 Cabinet.

Table C.3: Driver Details.

Diaphragm **Project: Sub Cabinet & Driver** **Variable and Value**	**Cone Front** **Description**
Meshing=Bifu	
DrvGroup=10001	Driving group link to observation stage
SubDomain=2	Front side belongs to sub-domain 3
Position=102,205,105 RefNodes="N"	Triangle giving position of Driver
//Simple=true	Simple model
Simple=false	Complex (detailed) model
dD=Di/1000	Diameter of cone+surr (337mm+11mm)
tD1=74/1000	Inner depth of cone to base of dust cap (mm)
dD1=75/1000	Diameter of dust cap (mm)
hD1=16/1000	Height of dust cap (mm)
hD2=95/1000	Height of total cone on the outside (mm)
dVC=50/1000	Diameter of voice coil (mm)
hVC=12/1000	Height of voice coil (mm)
dM=150/1000	Diameter of magnet (mm)
hM=20/1000	Height of magnet (mm)
Diaphragm	**Cone Rear**
RefDiaph="Cone front"	Inherite from "Cone front"
DrvGroup=10002	Driving group link to observation stage
SubDomain=1	This side belongs to sub-domain SubDomain=1
Side=rear	This is the rear side of the diaphragm
t1={-1*T/1000}//25.0mm	Cone Thickness

The loudspeaker driver details are shown in table C.3.

C.8   Running Our Model

We have now in theory produced our main model and we can mesh it as shown in figures C.8 and C.9.

Notice we are presented with two 'tick' boxes. These allow you to run the Observation Scripts immediately after the main model or not as your choice—this can be useful as sometimes you only need to run a particular output in a different way but have not changed the main model itself.

Ultimately we need some output from our work. This can be in two major forms: (1) A 3D field at the frequencies defined in the Solver section. (2) Various graphs or contours of the data for VACS is used to display the data and from which it can be exported to other programs.

It is important to realise that (1) also requires a mesh to be solved. It is tempting to set up a quick series of triangular or rectangular geometries and shapes perhaps overlapping the geometry of the model.

You could do this. However, if these shapes overlap different boundaries of the model, you will very likely get 'blank' spaces where no output solutions are displayed. There are at least two problems if this occurs.

Firstly, no field is shown in some areas so it is not clear if everything is working. Secondly, ABEC itself will be struggling to solve these areas. This also shows up as excessive time taken to produce the acoustic field display.

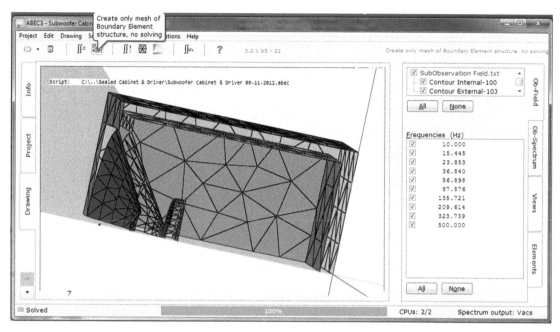

Figure C.8: Subwoofer Cabinet & Driver Meshing.

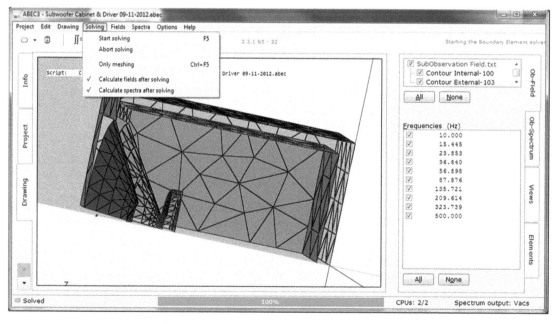

Figure C.9: Subwoofer Cabinet & Driver Solving.

In a test example, splitting the observation fields into separate non-overlapping regions increased the solving speed ten fold.

The reason being that this is a boundary element program and can only produce a valid output up to a boundary if a mesh triangle crosses from one domain into another. Then in that case, it has literally crossed the boundary and no solution is possible.

So if your model has gaps which do not appear to have solved, then check to see if there are elements of the observation mesh which are crossing boundaries and suffering from this problem.

The solution is quite simple, if a little long winded. What we need to do is to produce a series of individual mesh geometries that are completely inside any geometrical boundaries. This ensures that any Observation Meshes do not cross any boundaries. To be safe, it is advised to allow a safety margin or gap around the boundary edges.

The best way to do this is to add a variable representing a constant gap to the Formula section and then to produce a series of Observation Nodes. In this case we will save them as a separate file, 'SubFieldNodes.txt'. From this we can produce a series of observation planes to display the acoustic fields.

C.9 Acoustic Field Display (Observation Field)

As we said earlier, the Observation Field really needs its own nodes. We will set them up in the same way as the main nodes in a file 'SubFieldNodes.txt'. Also, again we will call the file 'SubWooferCabinetFormula.txt'.

```
Nodes N
Scale=0.001
Nodes "Contour"
  Scale=0.001

Field "Contour Internal"
  Meshing=Delaunay
  RefNodes="Contour"
  EdgeLength=20mm
  BodeType=LeveldB;  StepSize=1; Range=50 //BodeType=Phase
  Alpha=1
  100   301   300   302   303 // Side Plane Whole

Field "Contour External"
Meshing= Delaunay//Bifu
  RefNodes="Contour"
  EdgeLength=100mm
  BodeType=LeveldB;  StepSize=1; Range=50 //BodeType=Phase
  Alpha=1
  103   305   304   306   307
```

It is important before starting any analysis to have a clear idea of what output(s) you require and the level of detail or resolution required.

So in the case of our subwoofer, we only need information up to, say, 500 Hz. Probably a lower frequency would be acceptable but we might want to know the sound pressure distribution at various points in the 3D space. Also, we may need to know the SPL at 1 m and 2 m distances on axis as well as the electrical impedance as figure C.10.

There are two distinct requirements here: The 3D data or fields will be displayed by ABEC itself, but the other types of data will be displayed by a companion program VACS.

The observation field nodes are shown in table C.4.

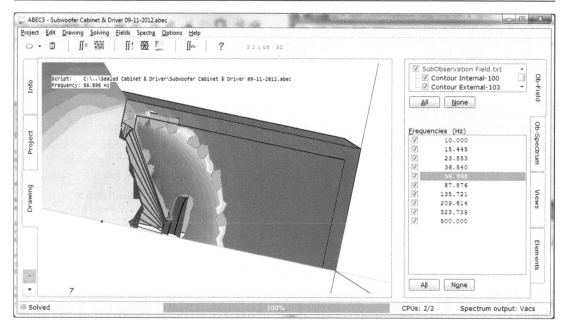

Figure C.10: Subwoofer Cabinet & Driver 57 Hz Field.

Table C.4: Observation Field Nodes.

NodeNo	X	Y	Z	Comments
300	0	0	0	//
301	0	0	D	//
302	0	H	0	//
303	0	H	D	//
304	0	H	0	//
305	0	H	D	//
306	0	H+1000	0	//
307	0	H+1000	D+1000	//

C.10   Acoustic Spectrum Display (SPL/Imp versus Frequency)

This requires a separate observation file to be produced: 'SubWooferDrvSpectrum.txt'. We will set them up in the same way as above and save them in a file calling both 'SubFieldNodes.txt' and 'SubWooferCabinetFormula.txt'.

```
Vacs_Project
  Name="Sub Woofer & Driver"
  VacsID=Sub Woofer & Driver
  NumFrequencies=200; Abscissa=log;

Nodes "Spectrum"
  Scale=0.001
  1001    0    {H+1000}  0  // Driver on-axis 1m
  1002    0    {H}       0 // Driver Near Field
```

```
BE_Spectrum
  PlotType=Curves
  RefNodes="Spectrum"
  GraphHeader="Sub Woofer & Driver_Sound Pressure"
  BodeType=LeveldB;  Range=50; //BodeType=LeveldB_Phase; Range=180

Name  Node
  101   1001    VacsID=SPL1m    // Driver On-axis outside  1m
  102   1002    VacsID=SPLDriverNearField//Driver Near Field

// Radiation_Impedance
//  GraphHeader="Sub Woofer & Driver_Radiation Impedance"
//  BodeType=Complex
//  RadImpType=Normalized
// Name DrvGroup DrvGroup
//  106  1001        1001    VacsID=ImpFrnt// Self front
//  107  1002        1002    VacsID=ImpRear// Self rear
//  108  1001        1002    VacsID=ImpMut // Mutual

LE_Spectrum "Sub Woofer & Driver_Impedance"
  AnalysisType=Impedance
  System="SubDrv"
  VacsID=Sub Woofer & Driver_Imp

LE_Spectrum
  GraphHeader="Sub Woofer & Driver_Velocity of VC"
  AnalysisType=Velocity
  System="SubDrv"
  109   "SubDrv"   VacsID=ImpVel_VC
```

Earlier we decided only to use ten frequencies for the Solver. This had the advantage of running very quickly but there are also consequences, as we shall see. Let's look at the spectrum or frequency response plots as produced by VACS, shown in figure C.11.

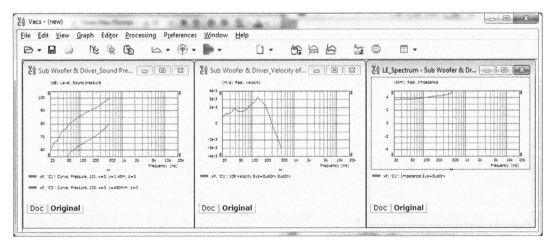

Figure C.11: Subwoofer Cabinet & Driver 1st VACS.

We can see a few problems:

- All the plots are over the 20 Hz to 20 kHz range but our model uses the 10 Hz to 500 Hz range—this is an automatic setting in VACS.
- The plots do not use the screens fully.
- The impedance plot is suspiciously flat, rising only to 4.6 Ω.

Lets take these in turn. First, the frequency range. Right click over the graphs as shown in figure C.12.

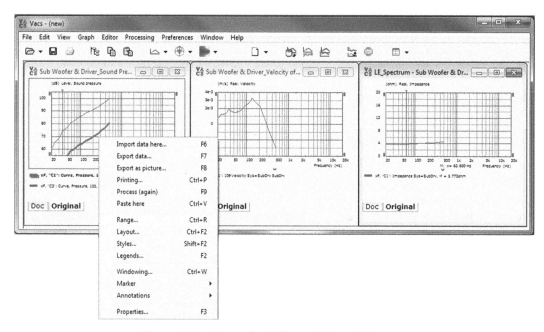

Figure C.12: Subwoofer Cabinet & Driver 2nd VACS.

Click Range as shown in figure C.13.

Figure C.13: VACS Graph Range.

We now need to go through the other graphs in turn; even so, the impedance curve as shown in figure C.14, is showing no detail.

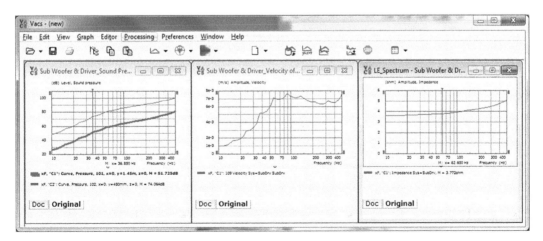

Figure C.14: Subwoofer Cabinet & Driver 3rd VACS.

What may be happening is that the main solution has insufficient detail in the frequency domain to accurately interpolate the data. Let's run it at 100 points to see if there is any difference in the results. This is often a good idea with FEA or BEM programs as they rely on interpolation.

Note

1. Although not strictly essential, we would *strongly* recommend using a Formula script as this can be used to pass common information to the other modules.

FEMM Tutorial

This chapter will build a detailed axisymmetric model of the magnetic circuit of the subwoofer motor unit, from constructing the geometry, setting up the boundaries assigning the materials, calculating the flux and looking for saturation, to estimating the voice-coil inductance.

D.1   David Meeker's Analysis of a Woofer Motor

FEMM, or Finite Element Method Magnetics, is a FEA magnetics program by David Meeker www.femm.info/wiki/HomePage.

Meeker has an analysis of a woofer motor on his website at www.femm.info/wiki/Woofer. We will base our analysis on much of his work but adapt it to our requirements.

D.2   Building an Initial Model

Our initial subwoofer driver earlier had the following details:

- 500 watts.
- 3-inch or 4-inch diameter voice coil in a 4-layer voice coil.
- X_{max} of 25 mm.
- Top plate thickness = 8 mm.
  - So wind length of 25 mm + 8 mm.

We can now search to see how close we can get to this with actual windings. We can we use the Theoretical $Bl(x)$ spreadsheet, or V-Coils, etc.

If we use 0.575 mm copper wire in a 4-layer with a 32 mm wind length using a 75 mm ID gives a DCR of 3.46 ohms; this has 208 turns and the OD = 77.5 mm. Our spreadsheet also says that for a 60 mm movement through the gap this would equal 100 turns in the gap, or for 48 mm movement this equals 80 turns in the gap.

So allowing 0.5 mm clearance inside and outside gives us a gap of around 78.0 mm–74.5 mm, or 3.5 mm.

We can now use FEMM to design the magnet structure. Let's try a 220 mm OD × 100 mm ID × 20 mm thick ferrite magnet. We can see the FEMM simulation of just the magnet by itself as figure D.1.

As we will be building an axisymmetric model, we will need the appropriate coordinates. It is usual in such a model to set the axis of symmetry to exactly zero. (Some programs get really confused if you do not do so. This can be especially difficult to ensure if you are importing from a program.)

For the axisymmetric model, we just need to use the radius rather than the diameter of parts and use the same vertical positions. In axisymmetric mode, FEMM uses a pair of coordinates: r-cord for the radius and z-cord for the

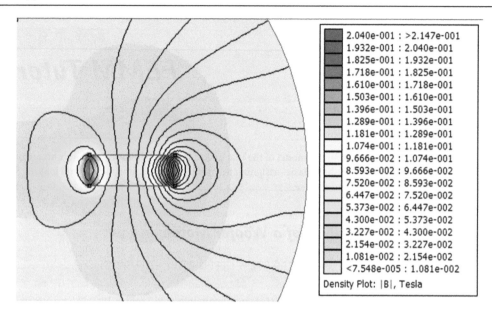

Figure D.1: Magnet Simulation.

height. Let us work out the coordinates for the various parts as follows: magnet = 220 mm outside diameter × 100 mm inside diameter × 20 mm thick.

The magnet will use 110 mm radius, 50 mm radius, and 20 mm thickness. So, the coordinates will be (110, −4), (50, −24), (50, −4), and (50, −24) for the magnet.

The back plate will use 105 mm radius and 8 mm thickness. So, the coordinates will be (105, −24), (105, −32), and (0, −32) for the back plate.

The top plate will use 105 mm radius, 39 mm radius, and 8 mm thickness. So, the coordinates will be (105, −4), (105, 4), (39, −4), and (39, 4) for the top plate.

The pole will use 37.25 mm radius, top at 4 mm, and bottom at −24 mm. The coordinates will therefore be (0, 4), (37.25, 4), and (37.25, −24) for the pole.

The voice coil will use 37.5 mm and 38.75 mm radius, top at 16 mm, and bottom at −16 mm. So, the coordinates will be (37.5, −16), (37.5, 16), (38.75, −16), and (38.75, 16) for the voice coil. To keep things simple, we have ignored the voice coil former. However, if this was made from a conductive or magnetic material this should also be included.

D.3   Setting up FEMM

Click on Problem and ensure it is set like this (as shown in figure D.2):

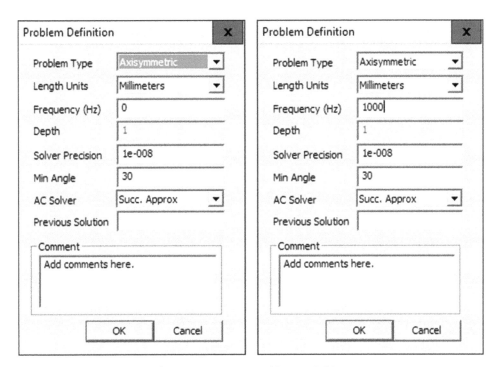

Figure D.2: FEMM Problem Definition.

Earlier, when we were discussing L_e and $L_e(x)$, and how this is dependent upon both displacement and frequency, we skipped over where we set up the model and so it's here.

If we wish to simulate the constant magnetic forcen the frequency (Hz) entry would be set to 0 Hz as here, but if we wish to simulate the A.C. component then change the frequency here to 100 Hz, 1000 Hz, or whatever is appropriate.

Then we will use our list of coordinates suitable for an axisymmetric model. Press TAB as shown in figure D.3.

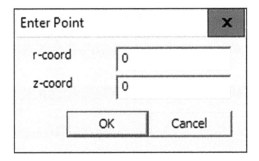

Figure D.3: FEMM Enter Points.

Enter 0,0 for the centre of a semi-circle. Then 0, 500 and 0, -500 for the top and bottom of the semi-circle (as shown in figure D.4):

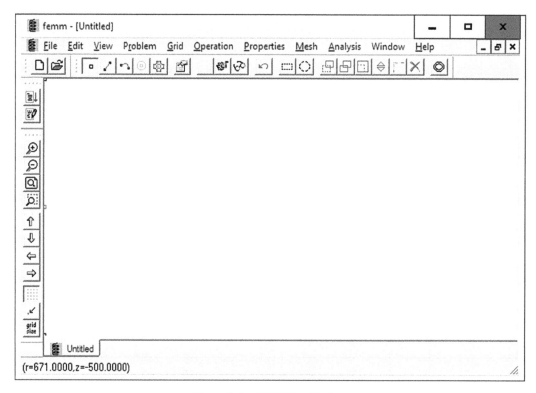

Figure D.4: FEMM 1st Nodes.

Click on the ARC Tool 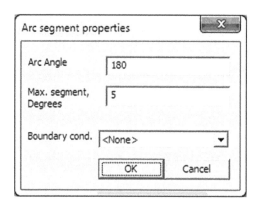. Click on the bottom point and then the top point as shown in figure D.5.

A dialogue box will open. Set it to Arc Angle = 180 & Max, segment = 5 and click OK.

Figure D.5: FEMM Arc Segment Properties.

We have now generated the main boundary we will use for our model. Select the Line Tool 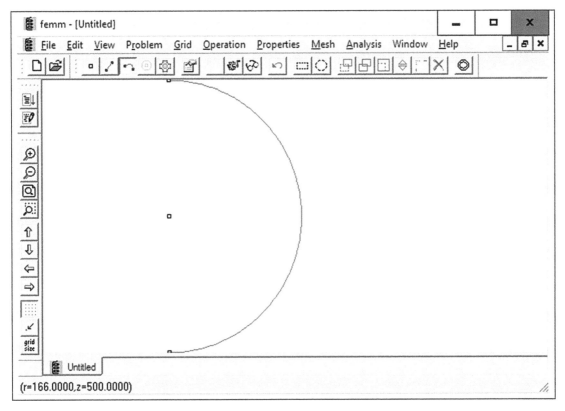 and click from the Top to the Centre and the Centre to the Bottom as shown in figure D.6.

Figure D.6: FEMM Arc Segment.

Select the Node Tool ⬚ . Press TAB enter and then 110, −4, then 50, −24, then 50, −4, thus generating the coordinates for the magnet as shown in figure D.7.

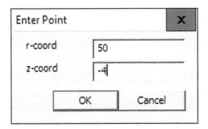

Figure D.7: FEMM Entering Points.

Save the file as FEMM Tutorial 1.

D.4 Drawing the Nodes

Select the Node Tool 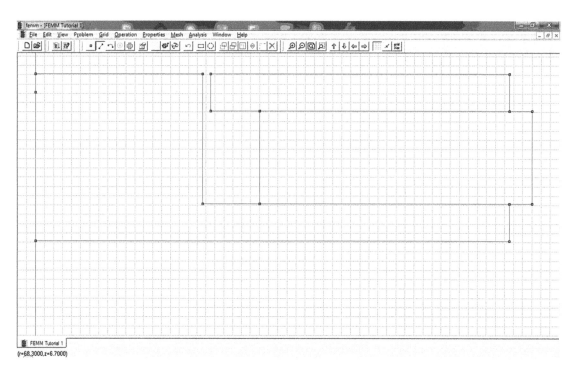 . Press TAB enter 105, −24 then 105, −32, and then 0, −32; this generates the Nodes for the Back Plate.

Select the Node Tool . Press TAB enter 105, −4 then 105, 4 and then 39, −4, and 39, 4; this generates the Nodes for the Top Plate.

Select the Node Tool . Press TAB enter 0, 4 then 37.25, 4, and then 37.25, −24; this generates the Nodes for the Pole.

Select the Line Tool . and click to fill in the Structure; it should look like this (as shown in figure D.8):

Figure D.8: FEMM Motor Unit Structure.

Save the Model.

D.5 Setting Boundary Conditions

Then we need to set up the boundaries for the model. Click Properties>Boundaries and then click Add Property as shown in figure D.9.

Property Definition
Property Name
New Boundary
Add Property
Delete Property
OK
Modify Property

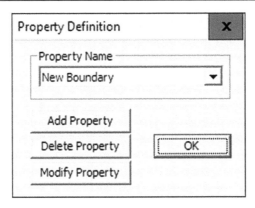

Figure D.9: FEMM Property Definition.

Replace the name New Boundary with ABC and change the BC Type to Mixed. The ABC name is meant to denote that we are creating an 'asymptotic boundary condition' that approximates the impedance of an unbounded, open space. In this way, we can model the field produced by the coil in an unbounded space while still only modelling a finite region of that space.

When the Mixed boundary condition type is selected, the c0 coefficient and c1 coefficient boxes will become enabled.

Select Properties and Boundary from the menu bar, then click the Add Property button. For our asymptotic boundary condition, we need to specify:

$$c_0 = \frac{1}{\mu_r \mu R} \tag{D.1}$$

$$c_1 = 0 \tag{D.2}$$

where R is the outer radius of a spherical problem domain. To enter these values into the dialogue box, enter 0 as the c1 coefficient and $1/(\mu 0 \cdot 500)$ as the c0 coefficient. FEMM includes the Lua scripting language which processes the contents of each edit box automatically when the dialogue is closed, substituting the numerical value of the permeability of free space for $\mu 0$ and evaluating the result as shown in figure D.10.

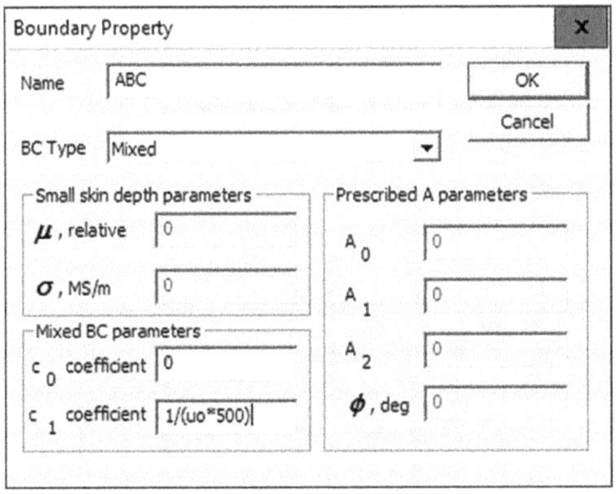

Boundary Property
Name
ABC
OK
Cancel
BC Type
Mixed
Small skin depth parameters
μ, relative
0
σ, MS/m
0
Mixed BC parameters
c 0 coefficient
0
c 1 coefficient
1/(uo*500)
Prescribed A parameters
A 0
0
A 1
0
A 2
0
φ, deg
0

Figure D.10: FEMM Boundary Property.

Press OK, just to check that it has worked correctly.

Click Modify Property. We can now see that c1 is now a number that is inversely proportional to the radius of the semi-circle boundary containing our model . . . OK

To assign this boundary condition, switch to 'operate' on arc segments mode. Select the arc defining the outer boundary by clicking on the arc with the left mouse button (the boundary will go RED as below), and push the space bar to open the arc's properties for editing. Select ABC from the boundary conditions drop list and click on OK. You have now defined enough boundary conditions to solve the problem, since a zero potential is automatically applied along the r = 0 line for axisymmetric problems as shown in figure D.11.

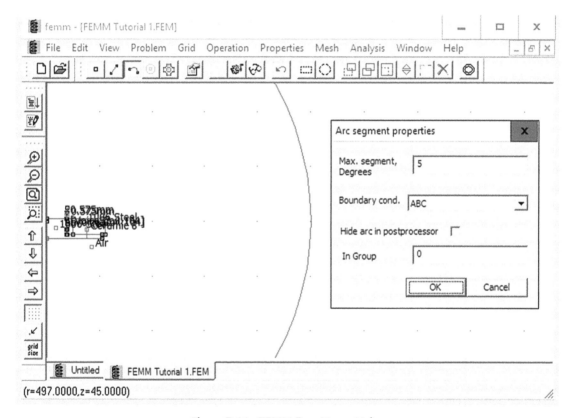

Figure D.11: FEMM Raw Motor Unit.

Press OK, just to check that it has worked correctly.[1]

D.6   *Selecting and Setting up Materials*

Click on Properties>Material Library.

Select and Drag the following into your Model Materials:

- Air.
- PM Materials->Ceramic Magnets->Ceramic 8.

- Soft Magnetic Materials->Low Carbon Steel->1006 Steel.
- Copper Metric Magnet Wire->0.575 mm.

Looking at the copper metric magnet wire, we cannot see a 0.575 mm wire—no problem, we can make one from 0.63 mm wire. Just click on material in the folder and right click to add new material, then change its parameters as required to produce 0.575 mm, then you can drag 0.575 mm to your model.

When you have all your materials click OK. Click the Material Selection Tool and click once in each component. Right click over each of these Material Nodes in turn and press the spacebar as shown in figure D.12.

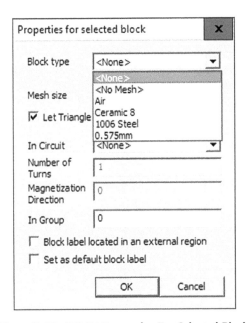

Figure D.12: FEMM Properties For Selected Block.

Under Block Type, select the appropriate material for each component. For the magnet, enter 90 for the direction of magnetisation as shown in figure D.13.

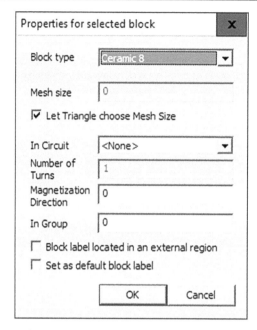

Figure D.13: FEMM Properties for Magnet.

Save the model.

D.7 Voice Coil

Select the Node Tool ⌐◻⌐. Press TAB enter 37.5, −16 then 37.5, 16, and then 38.75, −16, and 38.75, 16; this generates the nodes for the voice coil.

Then we need to draw the line through which we will check the flux simulation. but there is a problem, we wanted to check the simulation over a 48 mm length—clearly there is not sufficient depth to the back plate with this design.

Select the Node Tool ⌐◻⌐ .. Press TAB, enter 38.125, −24, then 38.125, 0, and lastly 38.125, 24. This generates the nodes for the simulation line; we then draw the measurement line. Now we need to apply the material to the voice coil, but first we need to make it a voice coil. So click Properties->Circuits.

Click Add Property. Change Name to Voice Coil and Current to 1 A. Click the Material Selection Tool ⌐◻⌐ . and click once in each component. Right click over each of these Material Nodes in turn and press the spacebar.

Our model now has all the essentials on it and should look like this:

D.8 Meshing Our Model

Click the Mesh Tool ▱.

D.9   Solving Our Model

Then Click the Solver ⬛. If the Mesher takes a long time (more than a couple of seconds), maybe there is a problem with the model. This can also show up as a very dense mesh where you do not expect it. The above mode is good, it is small, meshed, and it ran efficiently.

D.10   Viewing and Analysing the Results

Then the Viewer ⬛, you should then see something like this (as shown in figure D.14):

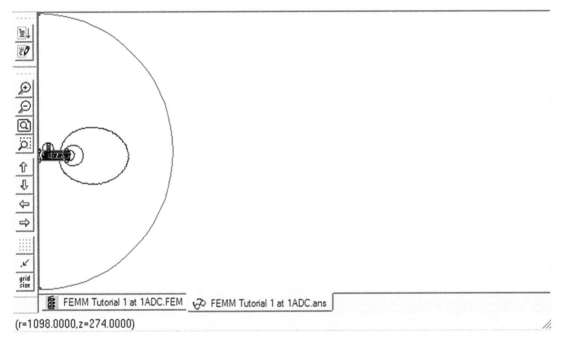

Figure D.14: FEMM Answer.

Great, but we cannot see anything! Our first job is to click the Zoom tool ⬛ and select around the model the region we are actually interested in (we would suggest the voice coil gap). Left click and form a box around the bits you wish to view (as shown in figure D.15):

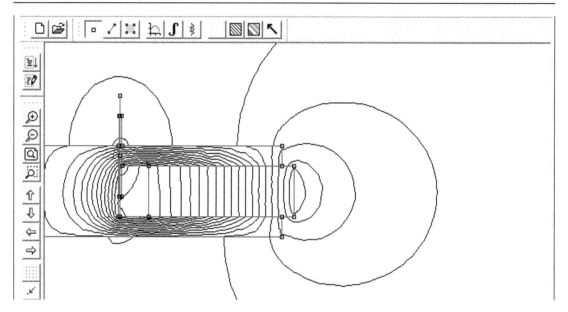

Figure D.15: FEMM Draw Zoom Box.

Click the Colour dialogue 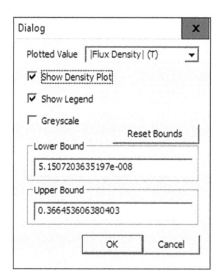 and tick Show Density Plot as shown in figure D.16.

Figure D.16: FEMM Density Plot Dialogue.

We can now see the flux density and saturation clearly as shown in figure D.17.

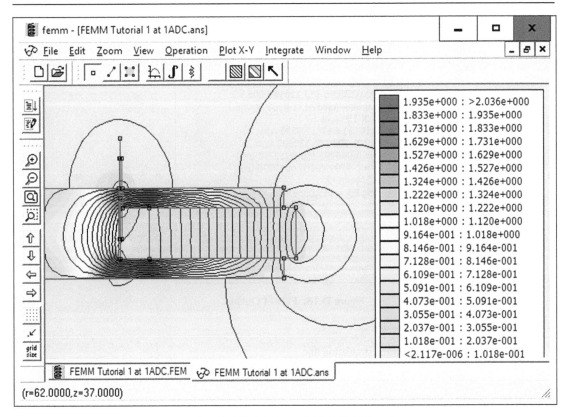

Figure D.17: FEMM Motor Unit 1 Field Density.

If we click on the Node tool ⌐ o ⌐ .and then on the Central Node we can read the flux at this position: FEMMOutput as shown in figure D.18.

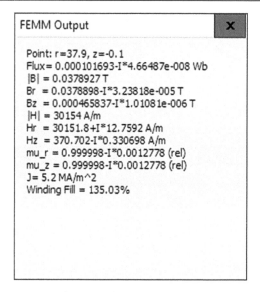

Figure D.18: FEMM Output.

Of more interest to us is the flux over the simulation line.

Click on the Line tool ⟋ then draw a line from the outside to inside.

D.11   Exporting Results

Click the Graph Tool ⌂. This shows figure D.19.

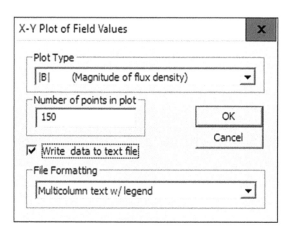

Figure D.19: FEMM X-Y Plot.

Save Result as FEMM Tutorial Backward 1 as shown in figure D.20. We can see in this case that we have set the number of points in plot to 150. This was to match the exact spacing of the voice coil wire gauge over the distance. When this is matched in the $Bl(x)$ spreadsheet we will later be able to predict the $Bl(x)$ of this magnet and voice coil combination.

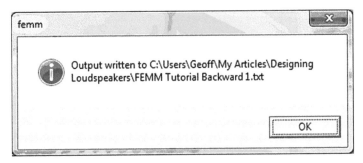

Figure D.20: FEMM Backward 1.

Click the Line tool again and press Escape. Click on the Line tool ✐ then draw a line from the inside to the outside. Click the Graph tool. 📈. This shows figure D.21.

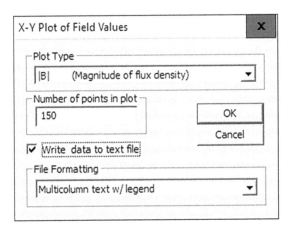

Figure D.21: FEMM X-Y Plot.

Save result as FEMM Tutorial Forward 1 as shown in figure D.22.

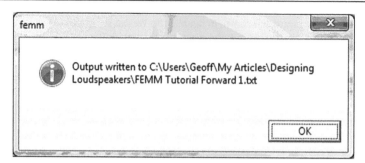

Figure D.22: FEMM Forward 1.

Alternately we could set up a Lua script to do the same thing—either way we can predict our $Bl(x)$ curve before we build anything!

D.12   Inductance

Firstly, it is essential that we update FEMM to the latest version[2] and open a copy of our first loudspeaker motor unit Subwoofer 1; this is shown as figure D.23.

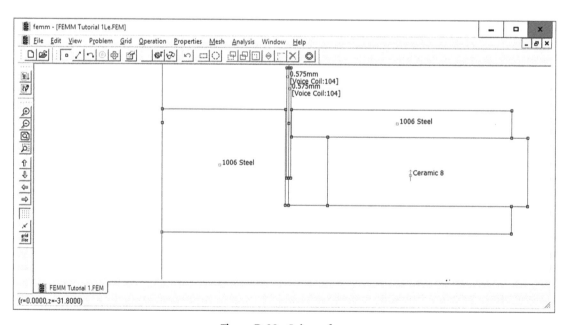

Figure D.23: Subwoofer.

Having loaded our model the first thing we *need* to do is to highlight the edges of the steel where we can expect the circulating currents to occur naturally. These will tend to be on the skin or the surface. We *right* click the lines as shown in figure D.24.

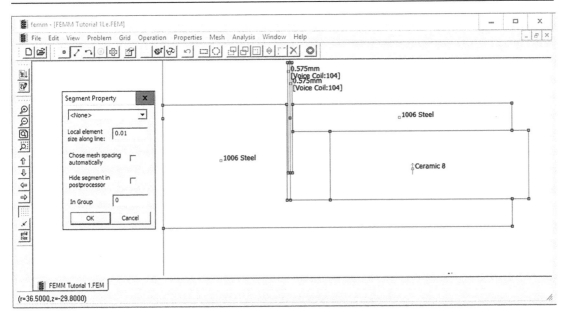

Figure D.24: Subwoofer 1 Select.

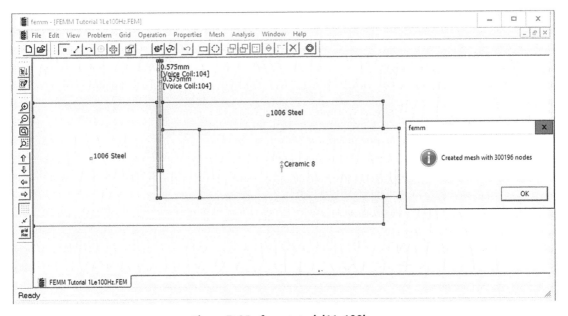

Figure D.25: femmtutorial11e100hz.

We then *uncheck* choose mesh spacing automatically—at this point we can enter in a mesh density to Local element size along line: for the moment we will enter 0.01.[3]

We will then set the model to 100 Hz and mesh it as shown in figure D.25, saving the result as shown in figure D.26.

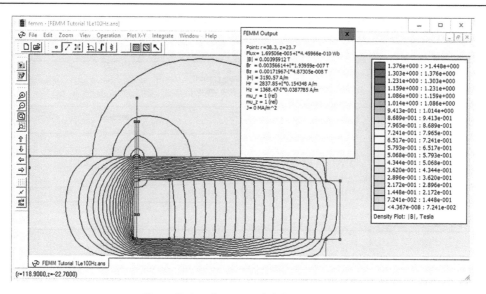

Figure D.26: femmtutorial1le100hzans.

As can be seen, there is now a much higher mesh density around the conductive lines we selected. Also, we have a much larger number of nodes: over 300,000, which will take a significant amount of time to solve. The resultant plot is shown as figure D.27.

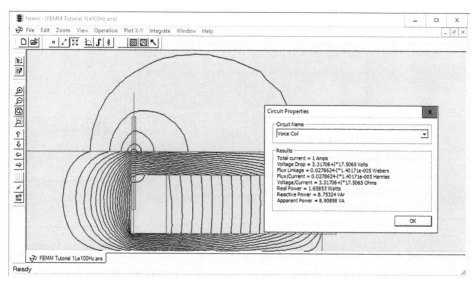

Figure D.27: femmtutorial1le100hzans.

Next we need to calculate the inductance. This is shown as figure D.27, where we read the voice coil circuit properties.

We can see here that the L_e at 100 Hz and with 1 amp calculates as 0.02786H or 28 mH—high, but possible: this would give an impedance of 17.5 ohms at 100 Hz.

However we also need to look at how this inductance varies with x as we saw in Chapter 20 on $L_e(x)$ on page 84.

Notes

1. Yes - we know it's very small—we'll get to that later . . .
2. Currently this is either femm 4.2 Nov 1 2015 (x64) or femm42bin_win32.
3. The skin depth at 20 kHz is 0.066 mm but we need to model several elements deep for sufficient accuracy, so we will use 0.01 mm.

Fusion 360 Tutorial

This tutorial uses Fusion 360[1] to produce and render a rotated 3D model of our subwoofer drive unit. Starting from initial sketches, these can form the core .dxf file, which could be used directly with other modelling programs, although Fusion 360 has significant inbuilt mechanical modelling capability. Mostly, however, these sketches will form 3D bodies, individual components, and our final 3D model.

E.1 Overview

Fusion 360 is available as an installable download for Windows and Mac platforms at relatively low cost (currently free for students and startups). By the time this edition is published it should also be fully accessible from a browser. Most importantly, you retain the rights to your work and ideas, and can freely export these models and files in a variety of open formats that can readily be used in other programs.

Fusion 360 is unusual in that it does not concentrate just on modelling, and more specifically on parametric modelling,[2] but rather it uses a 3D model to access a wide range of other functions. Being web/cloud-based, it is available to any user worldwide.

Heavy computational processing and calculation can be performed in the cloud rather than being completely dependent on your own individual computer. However, you do retain the option to solve and save files and models locally. Even quite complex models can be readily solved or visualisations rendered with relatively low-powered computers. Fusion 360 can be downloaded from www.autodesk.com/products/fusion-360/free-trial.

E.2 Fusion 360 Modules

As of February 2018, the following modules were available:

- MODEL: enables the construction of 3D Models or structures.
- PATCH: which has capabilities to handle 3D surfaces and geometries.
- SHEET METAL: which handles 2D sheets and folded structures.
- RENDER: where you can create produce photo realistic images.
- ANIMATION: which can show how a design is operated or assembled.
- SIMULATION: currently contains the following sub modules:
  - Static Stress.
  - Modal Frequencies.
  - Thermal.
  - Thermal Stress.
  - Structural Buckling.
  - Nonlinear Static Stress (Preview).
  - Event Simulation (Preview).
  - Shape Optimisation.

- CAM or Computer Aided Manufacturing, 2D and 3D tools for DRILLING, MULTI-AXIS, TURNING and CUTTING operations.
- DRAWING enables the production of conventional 2D drawings from the 3D Model.

Starting from sketching, the usual method is to draw 2D sketches in or on one of the 'normal' *X*, *Y* or *Z* planes. These are then refined or defined with a combination of dimensions and constraints to define or 'lock' them fully as sketches. You can then export these directly as .dxf files or extrude/revolve these 2D sketches into 3D parts.

E.3 Model

We find that for a loudspeaker drive unit, it is best to start out with an axisymmetric model, as this can quickly be revolved into a full 3D model. Let us start by selecting the model workspace as shown in figure E.1 by clicking on MODEL.

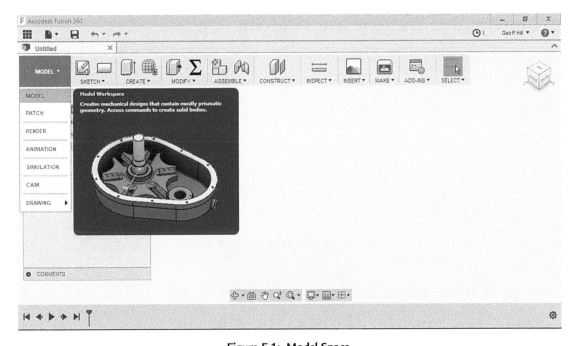

Figure E.1: Model Space.

Our first task should be to list out *all* of the components in the model (ideally with their sub components) that we wish to model. For our subwoofer this might be as follows:

- Back Plate and Pole
- Magnet
- Top Plate
- Voice Coil Assembly:
    - 0.575 mm Copper Wire
    - 0.1 mm Aluminium Former
    - 0.03 mm Kraft Paper
    - 0.5 mm Litz Braided Wire * 2 off

- Spider or Suspension
- Dust Cap
- Cone Surround Assembly: -
    - 0.2 mm Anodised Cone
    - 0.5 mm Nitrile Rubber Surround
- Chassis Assembly:
    - Chassis
    - M4 × 0.7, 7 mm Long Screws * 4 off
    - Terminal

Before starting, we recommend a little housekeeping to set up the interface, so that we can use the parametric and other capabilities a little more easily.

E.4   Change Parameters

Click on the dropdown arrow beside MODIFY, as shown in figure E.2.

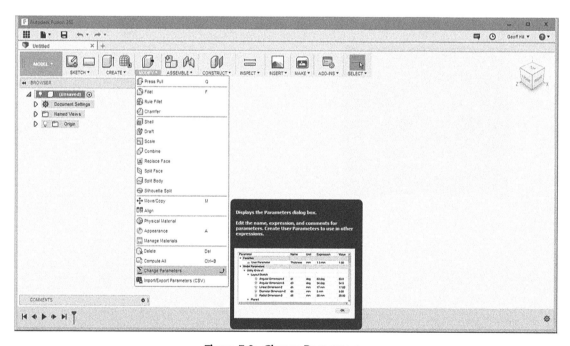

Figure E.2: Change Parameters.

Then click on the Add to Toolbar arrow to the right, as shown in figure E.3.

Figure E.3: Add Change Parameters to Toolbar.

This procedure can be used to add any of the possible selections to the toolbar.

Next we will go into the preferences; this is shown as figure E.4.

Figure E.4: Preferences.

Select Design and please ensure that Capture Design History is selected, as shown in figure E.5.

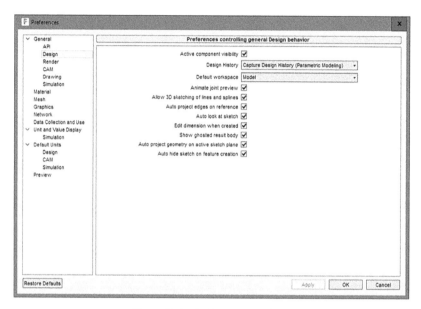

Figure E.5: Preferences Design.

E.5   Components

Let us now start the first component, the Back Plate and Pole (see figure E.6).

Figure E.6: Fusion 360 New Design.

Click New Design. This starts by opening a new Model, and displays a blank browser (see figure E.7).

Figure E.7: Fusion 360 New Browser.

We have highlighted the browser.[3] If you have a second screen available we strongly recommend locating it there, and not in the 'model' space. This is really important if you are building models with many parts or sub-assemblies, as this contains the functional parts list or bill of materials. Unfortunately, as of February 2018 the core of the structure of this parts list and the order of components cannot be easily changed, so it is useful to start it off correctly!

First, save this blank model, for example as 'Subwoofer Mk1'. The next task is to list all of the component parts, ideally including those parts included in sub-assemblies, such as the cone, surround, and voice coil. These can be entered later at any time; however, it is also easy to decide the best arrangement before starting the modelling/drawing process.

E.6   Building the Component List

To do this, right click on the name of your model, say, 'Subwoofer MKI'. This should be the only active part at the moment. The next job is to start naming the individual parts that comprise the subwoofer. We will start with the initial motor unit sub-assembly.

1. Backplate Pole
2. Magnet
3. Top Plate
4. Voice Coil Assembly
5. Spider or Suspension
6. Dust Cap
7. Cone Surround Assembly

So our first step should be to create these components in our model and then save our model as Subwoofer Mk1. The next task is to prepare a list of the main dimensions or sizes for these component parts, ideally with what you are going to call them.

We would suggest using simple, easily remembered names, for example a pole outside diameter might be PoleOD; a pole radius becomes PoleR, and so forth. This stage is not essential but will allow you to produce a parametric model where the relationship of the parts can be easily maintained and controlled, and changes simply made.

An example is shown as figure E.8.

BackPlateOD	210.00	mm	TopPlateThk	6.00	mm	CoilClr	0.30	mm	CoilOD	51.75	mm
BackPlateThk	6.00	mm	MagnetOD	220.00	mm	CoilWindLen	20.00	mm	TopPlateID	52.35	mm
PoleOD	50.00	mm	MagnetID	60.00	mm	PoleHt	26.00	mm			
TopPlateOD	210.00	mm	MagnetThk	20.00	mm	CoilID	50.60	mm			

Figure E.8: Subwoofer MKI Main Dimensions.

E.7 Using the Browser

Right click on the name of your model (in this case 'Subwoofer MK1') and select 'New Component' naming the first one 'Backplate' as shown in figure E.9. This component has been made active, as is shown by the dot to the right of the component in the browser.

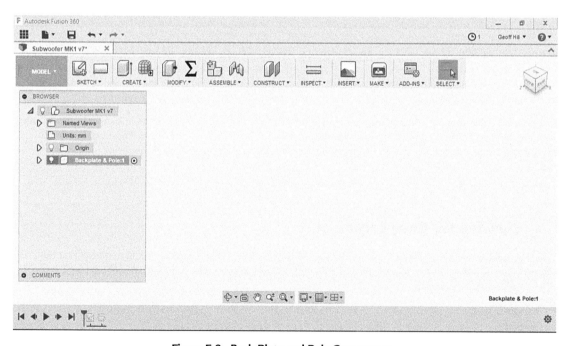

Figure E.9: Back Plate and Pole Component.

Unless you actually want to add a sub-level component to this one, you need to click back upon the main 'Subwoofer MKI' component and add further components from there. So click back on Subwoofer MKI and right click again, naming the next one as 'Magnet', and so forth until you have all the parts described in the model tree which should now look something like figure E.10. Note, however, that the voice coil, cone surround, and chassis are all assemblies, and clicking on these arrow(s) will expand them as desired.

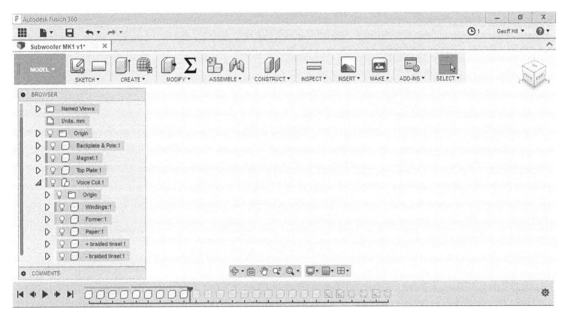

Figure E.10: Subwoofer Parts.

E.8   User Parameters

We recommend this optional step as it gives much greater flexibilty and ease of use and documentation. Click on the Σ symbol to the right of modify, to show the 'Add Parameters' dialogue as shown in figure E.11. Click onto the green + symbol. Enter the names and values you decided upon, as shown in figure E.8 on page 232.

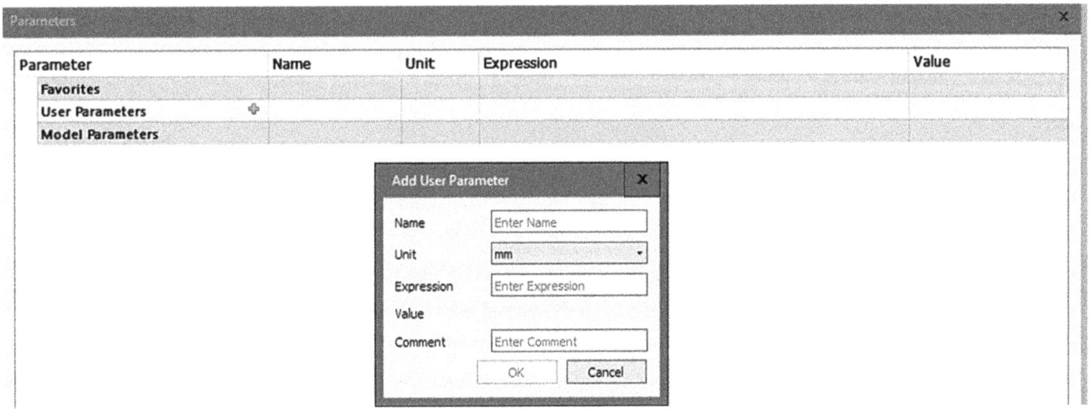

Figure E.11: Add User Parameters.

The initial user parameter table should like figure E.12.

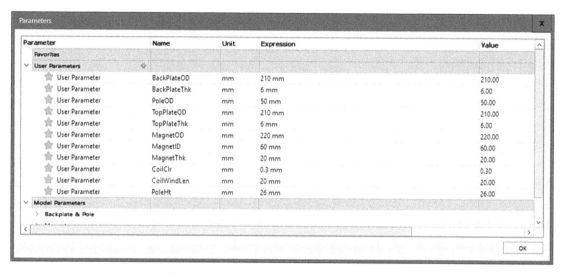

Parameter	Name	Unit	Expression	Value
Favorites				
∨ User Parameters ⊕				
⭐ User Parameter	BackPlateOD	mm	210 mm	210.00
⭐ User Parameter	BackPlateThk	mm	6 mm	6.00
⭐ User Parameter	PoleOD	mm	50 mm	50.00
⭐ User Parameter	TopPlateOD	mm	210 mm	210.00
⭐ User Parameter	TopPlateThk	mm	6 mm	6.00
⭐ User Parameter	MagnetOD	mm	220 mm	220.00
⭐ User Parameter	MagnetID	mm	60 mm	60.00
⭐ User Parameter	MagnetThk	mm	20 mm	20.00
⭐ User Parameter	CoilClr	mm	0.3 mm	0.30
⭐ User Parameter	CoilWindLen	mm	20 mm	20.00
⭐ User Parameter	PoleHt	mm	26 mm	26.00
∨ Model Parameters				
> Backplate & Pole				

Figure E.12: Subwoofer MKI User Parameters.

Notice that here we have just listed a name and individual dimensions for various aspects of these parts; later we will see how we can use these *Names* effectively.

E.9   Start Modelling

Now we can start to sketch the individual components in turn and thus build our full model. We will click on the Backplate and Pole to activate this component. Then clicking on 'Sketch' shows the blank 'Origin' and the three default planes centred on $X = 0$, $Y = 0$, and $Z = 0$ as shown in figure E.13.

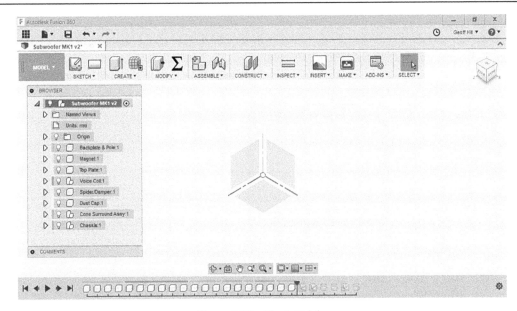

Figure E.13: Blank Origin.

Notice that you can click on any or indeed all of the three planes and highlight them without anything else happening until you click on 'Create Sketch' at this point. First, the background grid orientates to the selected plane; then when you click on any of the planes the screen re-orientates to that plane. For example, the X (red) and Y (green) are shown as figure E.14.

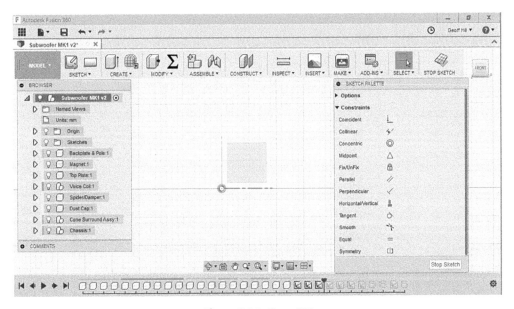

Figure E.14: Start *XY*.

Notice that the 'sketch pallette' sits to the right of the origin. This contains options which we will leave at the default settings, and constraints which we will need shortly. Under 'sketch', click Line or press L, as shown in figure E.15. You can then use the cursor to 'draw' the shape you require.

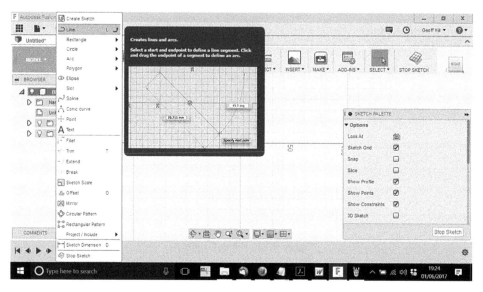

Figure E.15: Start Sketch.

Starting from the origin, click on this and move the cursor vertically or at 90° for a distance of around 40 mm notice the *Horizontal/Vertical constraint* has been applied automatically[4] as we moved at 90°. This is shown as figure E.16. You can be precise with your dimensions at this point if you wish, but it is not essential as we will be making this a parametric model.

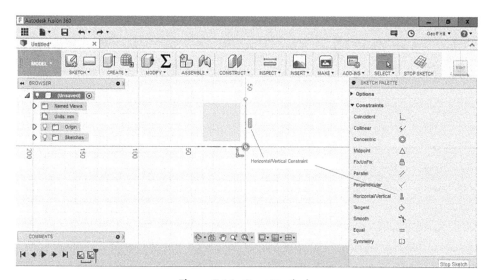

Figure E.16: Start Vertical.

Next we 'draw' a line horizontally approximately 10 or 15 mm. Notice again the *perpendicular constraint* has been applied; this is shown as figure E.17.

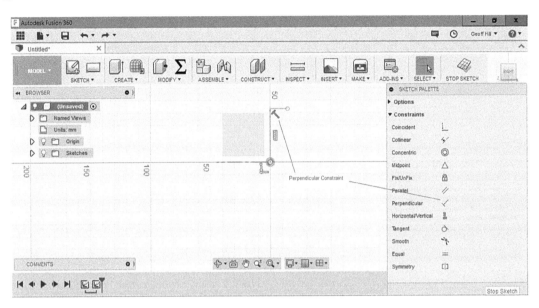

Figure E.17: Start Horizontal.

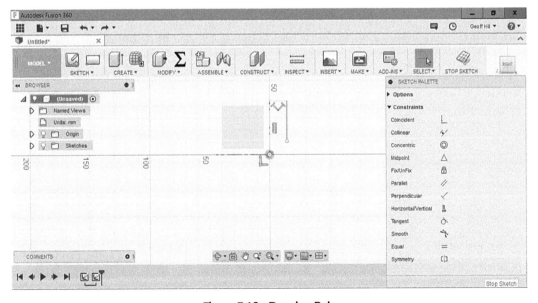

Figure E.18: Drawing Pole.

We continue drawing individual lines (again applying constraints as we go) until we have the outline of the Backplate and Pole as shown in figures E.18 and E.19. Notice that we have not used any dimensions yet.

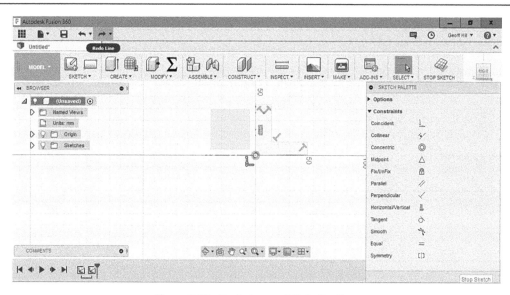

Figure E.19: Backplate and Pole Outline.

Clicking on the top of the Pole, we have box where we can enter a dimension directly as shown in figure E.20.

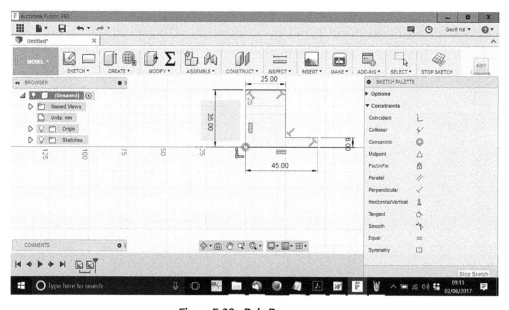

Figure E.20: Pole Parameters.

Clicking on the top of the Pole, we find a box where we can enter a dimension directly. However, if instead we enter 'P', this will highlight any of the user parameters we entered earlier with 'P' in them. We will select PoleOD, and as this is a radius, enter it as PoleOD/2, as shown in figure as shown in figure E.21.

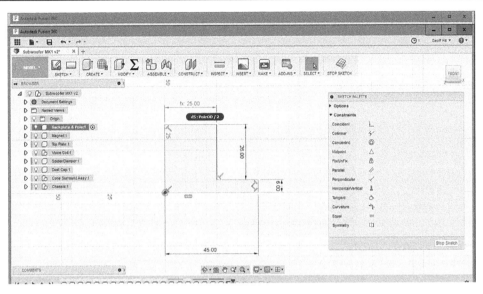

Figure E.21: Pole User Parameters.

We can then continue applying these user parameters to each of the individual dimensions in turn, on an individual component or indeed the whole assembly. We will finish this section with the sketch of the Pole, fully dimensioned as shown in figure E.22.

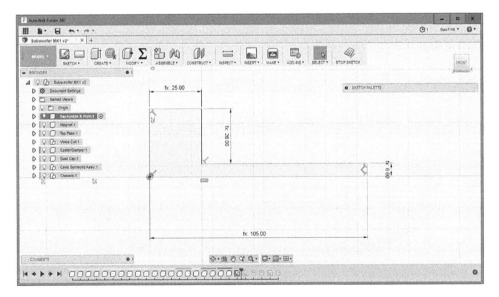

Figure E.22: Pole Final Sketch.

Click top 'sketch' to finalise this and start the process of revolving this sketch into a 3D model as shown in figure E.23.

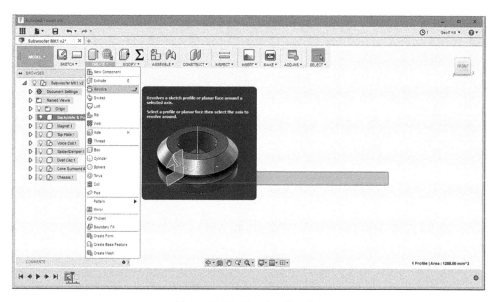

Figure E.23: Revolve Sketch.

Tilting the model slightly we can see the full 360° of the Pole, as shown in figure E.24.

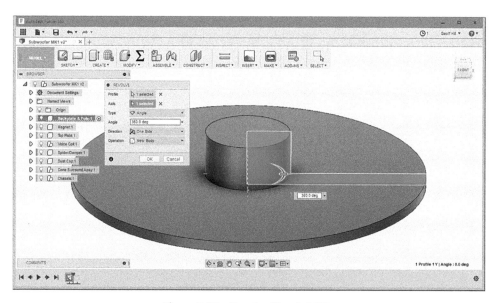

Figure E.24: Revolve Sketch 360°.

We can now continue a similar process with all the other parts.

E.10   Magnet

So, for example, the Magnet sketch should be as shown in figure E.25.

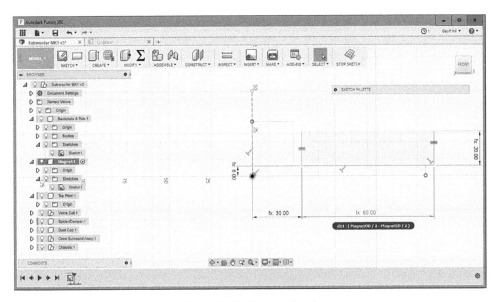

Figure E.25: Magnet Sketch.

We then need to revolve this sketch around the same *Y*-axis as used for the Backplate and Pole previously, as shown in figure E.26.

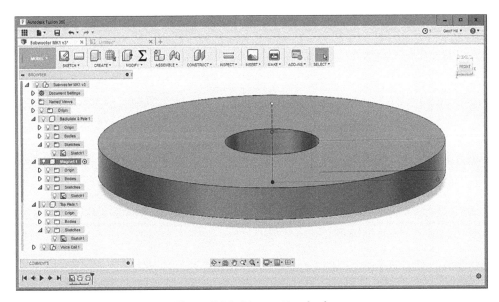

Figure E.26: Magnet Revolved.

E.11   Top Plate

The Top Plate sketch is shown in figure E.27.

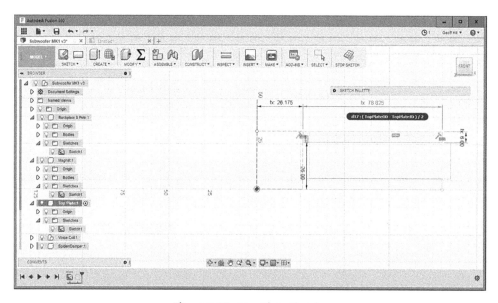

Figure E.27: Top Plate Sketch.

We then need to revolve this sketch around the same Y-axis as used for the Backplate, Pole, and Magnet previously, as shown in figure E.28.

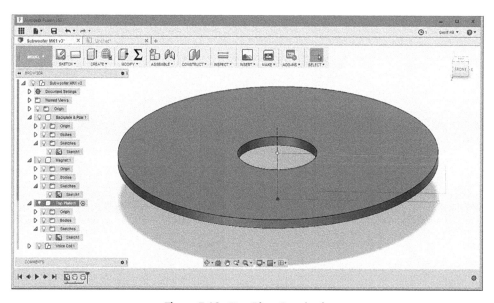

Figure E.28: Top Plate Revolved.

E.12   *Voice Coil Assembly*

The voice coil is of course made up from several separate components:

E.12.1   *Former*

The Former sketch is shown in figure E.29.

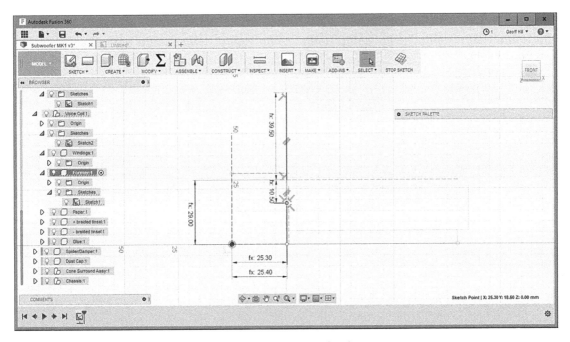

Figure E.29: Former Sketch.

We then need to revolve this sketch around the same Y-axis as previously, as shown in figure E.30.

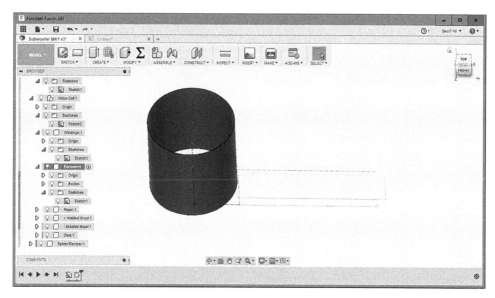

Figure E.30: Former Revolved.

E.12.2   Windings

The voice coil windings sketch is shown in figure E.31.

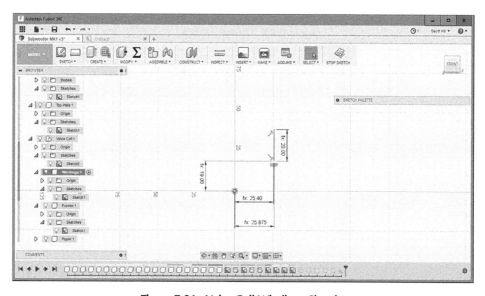

Figure E.31: Voice Coil Windings Sketch.

We then need to revolve this sketch around the same *Y*-axis as previously, as shown in figure E.32.

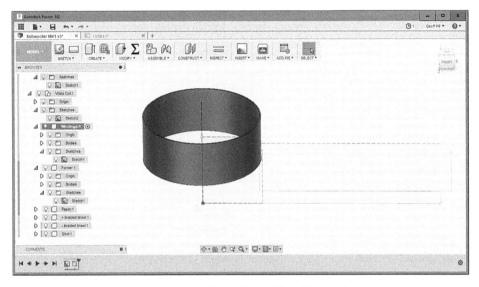

Figure E.32: Voice Coil Windings Revolved.

E.13 Initial Motor Unit Full Assembly

The full motor unit assembly is as shown in figure E.33. To help see the individual parts clearly, we have applied the 'Component Color Cycling Toggle', which can be found under the 'Inspect' menu, and can also be accessed by *Shift + N*.

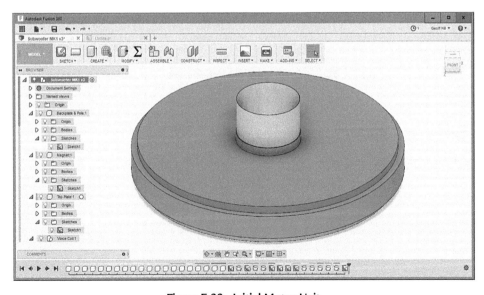

Figure E.33: Initial Motor Unit.

Comparing this initial motor unit with those from earlier, we can see that many of the sizes are not what we might wish. This is no problem, as we only need to update our user parameters as shown in figure E.34 and let Fusion 360 update the full model.

Parameter	Name	Unit	Expression	Value	
Favorites					
⌄ **User Parameters** ✚					
☆ User Parameter	BackPlateOD	mm	210 mm	210.00	
☆ User Parameter	BackPlateThk	mm	6 mm	6.00	
☆ User Parameter	PoleOD	mm	74.4 mm	74.40	
☆ User Parameter	TopPlateOD	mm	BackPlateOD	210.00	
☆ User Parameter	TopPlateThk	mm	6 mm	6.00	
☆ User Parameter	MagnetOD	mm	220 mm	220.00	
☆ User Parameter	MagnetID	mm	80 mm	80.00	
☆ User Parameter	MagnetThk	mm	20 mm	20.00	
☆ User Parameter	CoilClr	mm	0.3 mm	0.30	
☆ User Parameter	CoilWindLen	mm	20 mm	20.00	
☆ User Parameter	PoleHt	mm	MagnetThk + TopPlateThk	26.00	
☆ User Parameter	CoilID	mm	PoleOD + (2 * CoilClr)	75.00	
☆ User Parameter	CoilOD	mm	77.5 mm	77.50	
☆ User Parameter	TopPlateID	mm	CoilOD + (2 * CoilClr)	78.10	
☆ User Parameter	FormerLength	mm	30 mm	30.00	
☆ User Parameter	ConeOutsideRadius	mm	158.8 mm	158.80	
☆ User Parameter	ConeHeight	mm	108.3 mm	108.30	
☆ User Parameter	Revolve	deg	360 deg	360.0	
☆ User Parameter	ConeDepth	mm	64.3 mm	64.30	
☆ User Parameter	SurroundRad	mm	5 mm	5.00	
☆ User Parameter	SurroundThk	mm	0.5 mm	0.50	
☆ User Parameter	conethk	mm	0.3 mm	0.30	
☆ User Parameter	coneneck	mm	3 mm	3.00	
☆ User Parameter	ChassisStep	mm	9.25 mm	9.25	
☆ User Parameter	RollLength	mm	46.522 mm	46.522	
⌄ **Model Parameters**					

Figure E.34: Updated Parameters.

E.14   *Spider or Damper*

An initial rendering is shown in figure E.35.

Figure E.35: MK1 Motor Unit and Chassis.

Then we need to locate the Spider or damper, we will start with an axisymmetric sketch as shown in figure E.36.

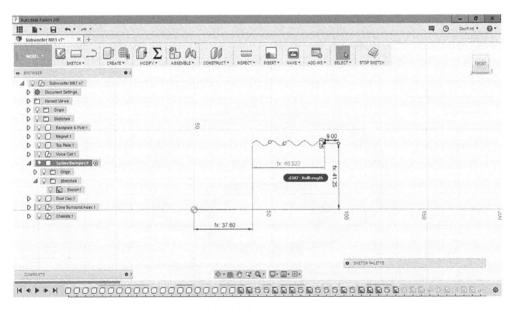

Figure E.36: Spider or Damper Sketch.

Next we take this profile and revolve around the central Y-axis, as shown in figure E.37.

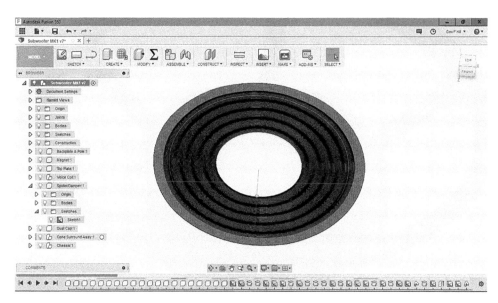

Figure E.37: Spider or Damper Revolved.

This needs to be positioned within the chassis, shown in figure E.38.

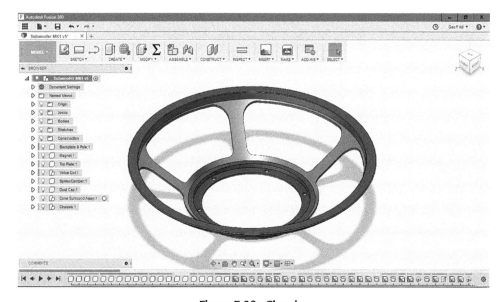

Figure E.38: Chassis.

E.15 Final Assembly

This is shown in figure E.39 and a fully rendered version in figure E.40.

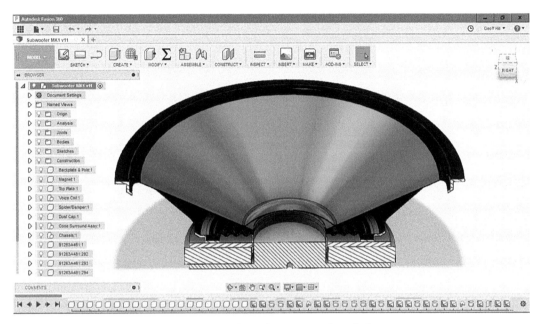

Figure E.39: Subwoofer MK1 Sectioned View.

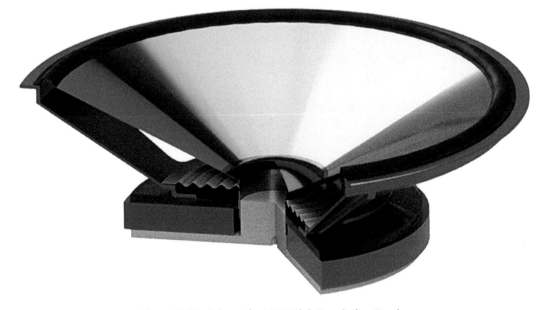

Figure E.40: Subwoofer MK1 High Resolution Render.

Notes

1. From Autodesk.
2. Parametric modelling is where a model is designed and developed with reference to other dimensions or parts.
3. The position of this browser is *not* defined and it can appear anywhere on the screen.
4. You can apply or remove any constraint manually

HOLMImpulse Tutorial

We use HOLMImpulse to demonstrate the process of measuring and calibrating a Tetrahedral Test Chamber. HOLMImpulse is a freeware program for measuring the Impulse, Amplitude including THD+n, Harmonics, and Phase versus Frequency.

HOLMImpulse lets you select THD, Noise, 2nd HD, 3rd HD, 4th HD, 5th HD, 6th HD, 7th HD, 8th HD, 9th HD, and 10th HD as well as Phase. By default it works using a dB or Logarithmic Y axis. However, this can be changed for impedance measurements.

F.1   Initial Setup

Open HOLMImpulse and click on Device & Signal.

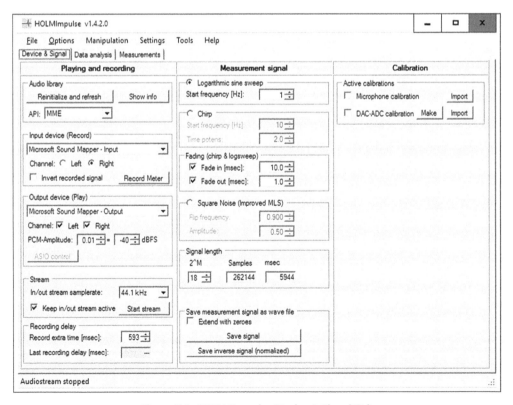

Figure F.1: HOLMImpulse Device & Signal Tab.

Without a sound card connected, the Device & Signal Tab should look like figure F.1.

Notice that the audiostream is stopped. The first thing to do is to connect the sound card—in this case a Rowland UA-25EX—then click 'Reintialize and refresh' and then we can select the UA-25EX for the input and or output. Then change the in/out stream sample rate as appropriate and click 'Retart stream' as shown as figure F.2.

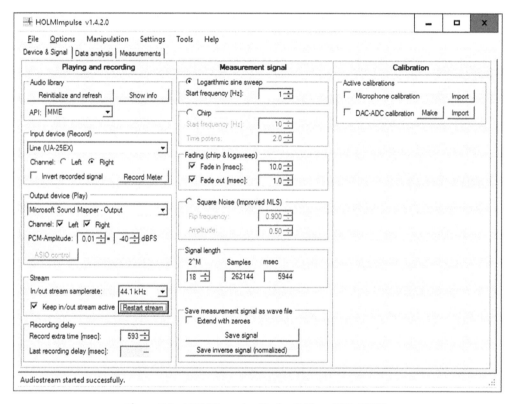

Figure F.2: HOLMImpulse Device & Signal UA-25EX.

If you have a calibrated microphone, you can enter its information under 'Calibration'.[1] In practice, with a Tetrahedral Test Chamber I recommend leaving the microphone correction unchecked and unused and following the procedure below. Also, you can fairly safely leave the DAC-ADC calibration unchecked as well, as this will also be taken into account at the same time.

The measurements here will be using the UA-25EX in full duplex mode in which this sound card is limited to a maximum sample rate of 48 kHz.[2]

F.2   Internal Measurement in a Tetrahedral Test Chamber

This is shown as figure F.3.

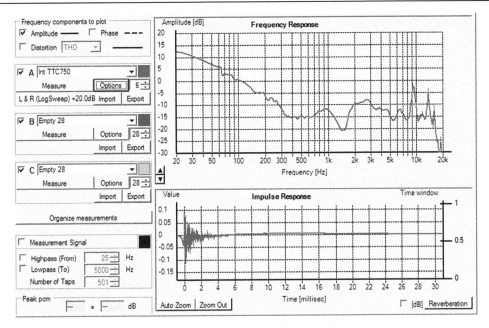

Figure F.3: SEAS H1207 inside a TTC750.

Next we need to click the Options button next to 'Measure'; this is shown as figure F.4.

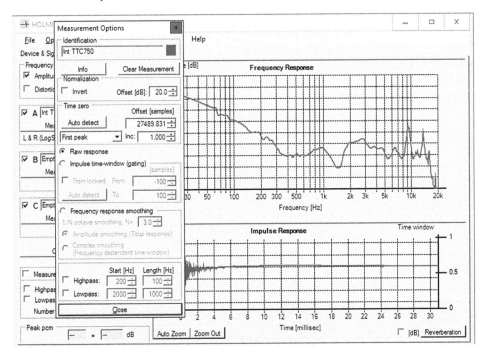

Figure F.4: TTC750 Internal Options.

F.3 External Measurement in a Tetrahedral Test Chamber

This is shown as figure F.5.

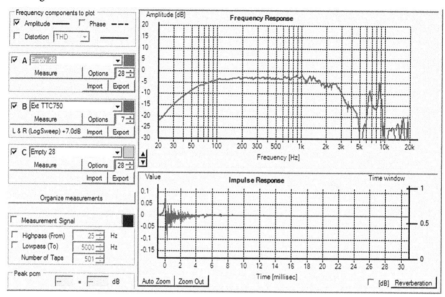

Figure F.5: SEAS H1207 Very Near Field Outside a TTC750.

F.4 Making a Correction Curve

This is shown as figure F.6.

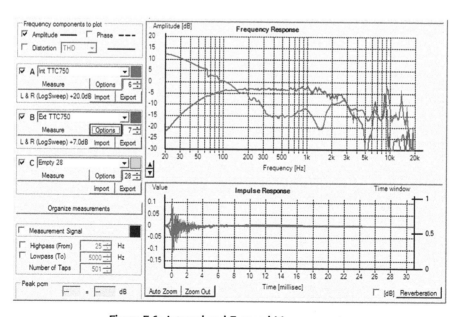

Figure F.6: Internal and External Measurements.

Next we need to click on Manipulation->Division as shown in figure F.7.

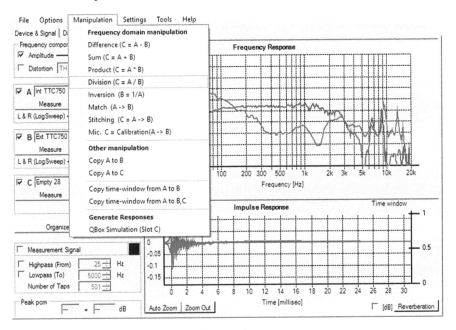

Figure F.7: Division of Measurement.

The result is shown as figure F.8.

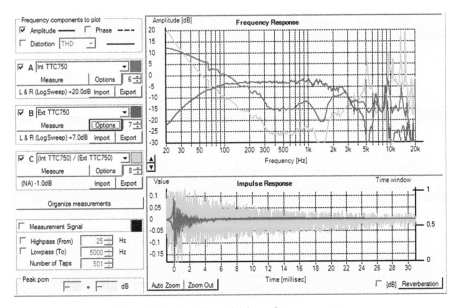

Figure F.8: Resulting Division of Measurement.

We clearly have a reasonably good equalisation curve at low frequencies, but it is not valid at high frequencies.

The frequency to which the correction is valid will depend upon the driver used to make these measurements. In the case of the SEAS H1207 used here it should be reasonably *pistonic* to around 1700 Hz.

To do this we need to *Stitch* this rough correction curve to a 0 dB curve above 1700 Hz.

We make a 0 dB curve by dividing any curve by itself, as shown in figure F.9.

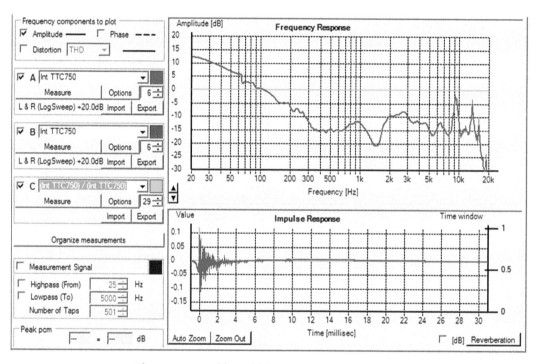

Figure F.9: Resulting Division of Same Measurement.

The green line is at 0 dB; we *Stitch* our rough correction curve to this from 1700 Hz.

First though we highlight the name in the C Measurement title and change to 0 dB.

We select a rough equalisation curve in A and the 0 dB curve in B with an empty slot as C.

This is shown as figure F.10.

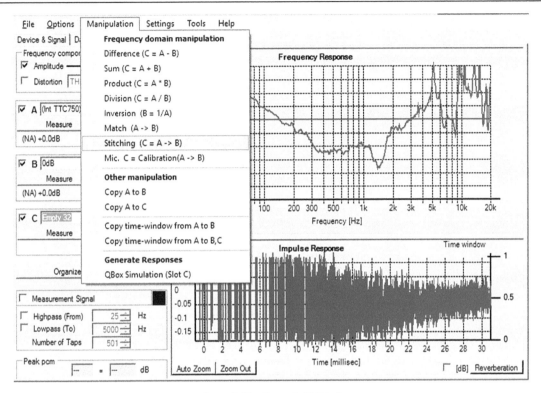

Figure F.10: Eq to 0dB Stitch.

When we select *Stitching (C = A -> B)*, a dialogue with Frequency Options appears and we enter 1700 Hz & 50 Hz. We then click Update.

This is shown as figure F.11.

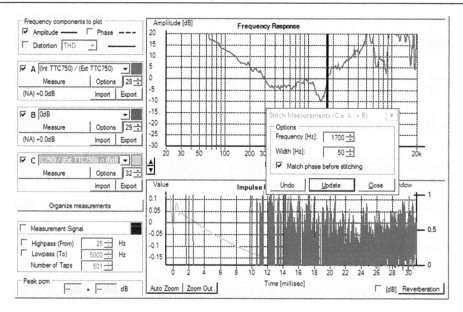

Figure F.11: Eq to 0dB Stitch Option.

The curve has moved up, the final equalisation curve is shown as figure F.12.

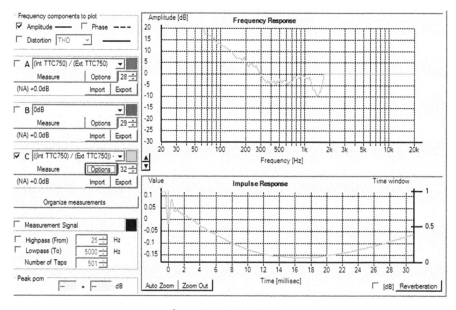

Figure F.12: TTC750 Eq.

We would suggest naming this equalisation curve immediately in this case as TTC750EQ.

F.5   *Applying the Correction Curve*

We then select this equalisation curve in B with the internal measurement in A and click Manipulate, as shown in figure F.13.

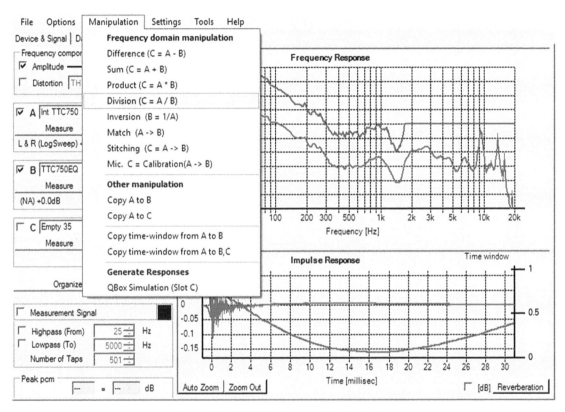

Figure F.13: TTC750 Internal Divided by Eq.

Click Divide and the result is shown as figure F.14.

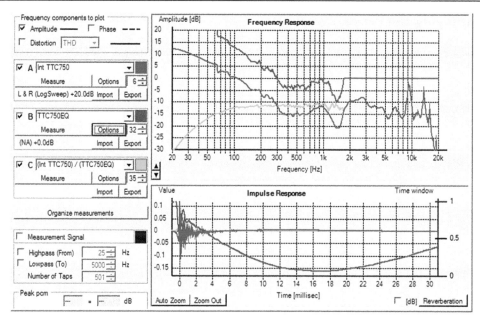

Figure F.14: TTC750 Result.

F.6   Final Result

We have unchecked the other curves and selected this result, naming it TTC750 Final Result in figure F.15.

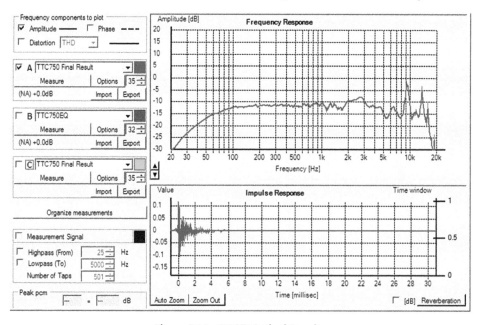

Figure F.15: TTC750 Final Result.

F.7   Impedance Measurements

Impedance measurements are possible using HOLMImpulse but are slightly more complicated. Firstly, we need to decide which method for measuring impedance to use and make up the appropriate connectors. Secondly, we would recommend having a few known resistor values available to calibrate the measurements against. In this case we would recommend using a 1 ohm and 6 ohm resistor as a minimum, and the high impedance method, where we will use a series resistor of 800 ohms.

The circuit is shown in figure F.16.

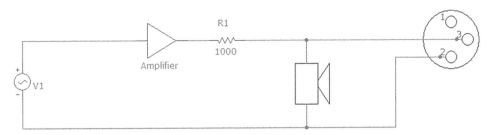

Figure F.16: Measuring Impedance using High Impedance Method.

The first task is to measure the voltage across both the 6 ohm and 1 ohm resistors via the 800 ohm source resistor, as shown in figure F.17.

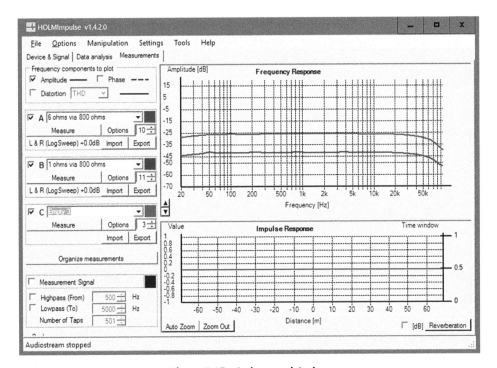

Figure F.17: 6 ohms and 1 ohm.

The next step is to divide the 6 ohm measurement by the 1 ohm measurement: Manipulation->Division (C = A/B) as shown in figure F.18.

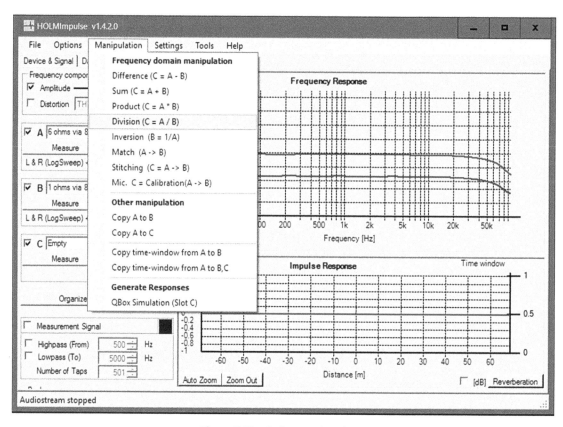

Figure F.18: 6 ohms and 1 ohm.

The result of this division is shown in figure F.19.

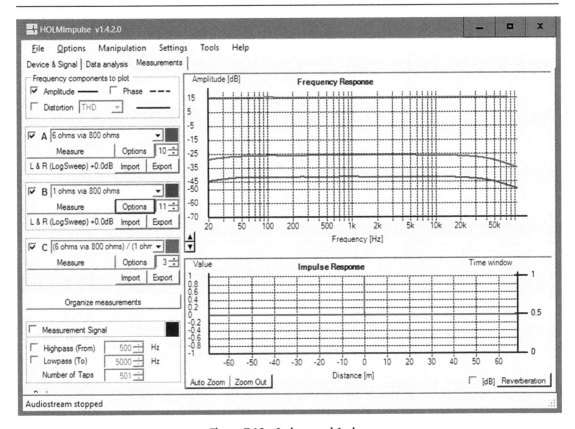

Figure F.19: 6 ohms and 1 ohm.

We then need to convert between dB scale and ohms. To do this we need to change the vertical (Y) scale to linear: Options->Linear amplitude axis, as shown in figure F.20.

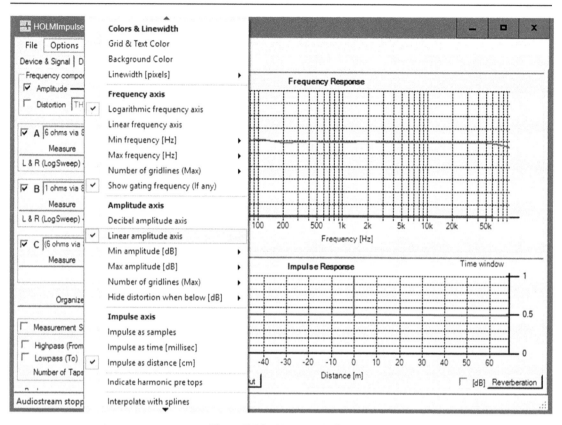

Figure F.20: Linear Y Scale.

The final result confirming measurement and calibration is shown in figure F.21.

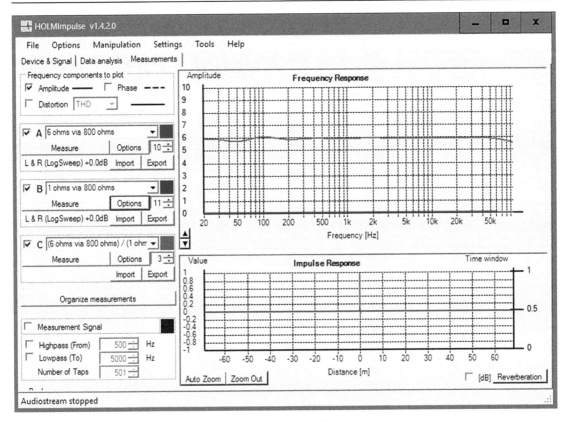

Figure F.21: 6 ohms Linear.

Notes

1. The range of correction here is severely limited to around 12 dB in total.
2. Full duplex as in sending and receiving simultaneously.

Klippel LPM Tutorial

To demonstrate the process of measuring the parameters of a loudspeaker using the Klippel System, we use the Klippel example database so you can follow without the hardware. The actual measurement was conducted by the LPM (Linear Parameter Module), one of the optional modules of the Klippel R&D analyser.[1]

We will assume that you are have a Klippel R&D system up and running. If not, you can follow this tutorial to gain insight into LPM measurement and accessing data without making a measurement.

In the Klippel system, any measurement or simulation is contained within an individual database object. This object contains the results of the measurement or simulation as well as all the individual set-up information necessary for the given measurement or simulation to run, provided it is associated correctly with the required hardware, software dongle, or licences within the Klippel dB-Lab framework.

The Klippel R&D and QC software viewers can be freely downloaded; this allows for measured data to be exchanged between users and for us to examine typical results.

We will start with a blank R&D System as shown in figure G.1. Click on the blue database icon and select an individual database.

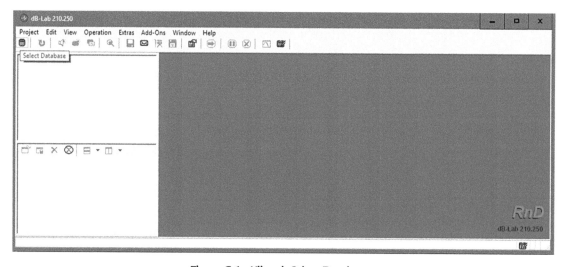

Figure G.1: Klippel: Select Database.

G.1 dB-Lab

dB-Lab is the database framework which supports and organises all of the measurements or simulations undertaken in the R&D, QC, Power Test (PWT), Material Parameter Measurement (MPM), or Scanning Vibrometer (SCN).

We will use the R&D example database and select the Transducer Parameter Identification (LSI, LPM, MMT, MSC, BAC, IMP) as shown in figure G.2.

We then select the Multimedia Woofer (LPM, LSI) and select the first database 1 LPM T/S parameter with laser, as shown in figure G.3.

We covered the Small Signal Theory in chapter 7 on page 37 When we select any of the database objects available, this opens an individual database within the dB-Lab framework, as shown in figure G.4. There are two key regions within this; to the left and top is the *File Browser*, whilst to the left and the bottom is the *Measurement Results Selection*.

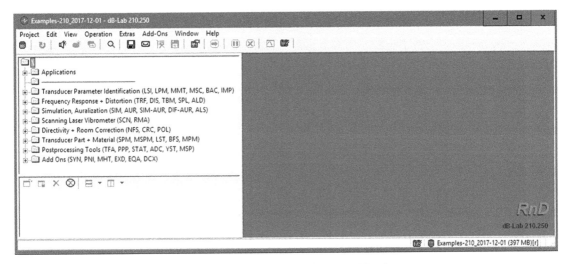

Figure G.2: Klippel - Transducer Identification.

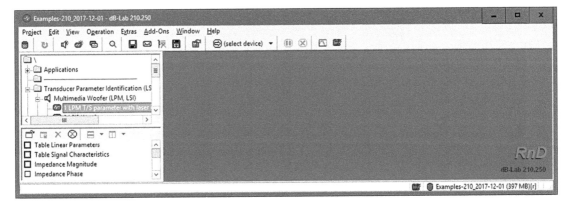

Figure G.3: Klippel: Select LPM.

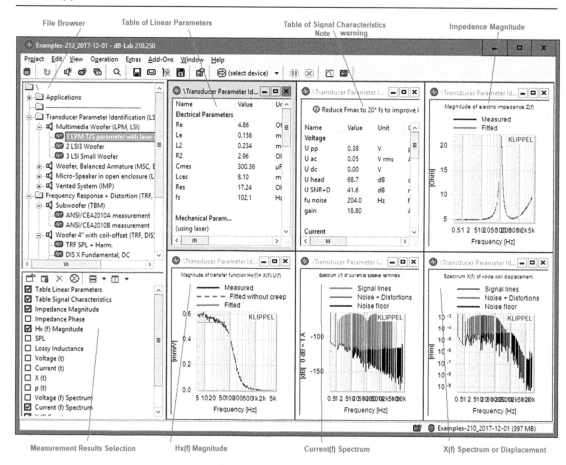

Figure G.4: Linear Parameter Measurement (LPM).

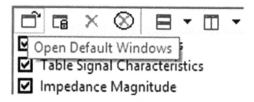

Figure G.5: Open Default Windows.

Considering the Measurement Results section first, there are five icons at the top of this section that you need to know about as they are the key to displaying the relevant information about any individual measurement or a series of measurements. These are as follows.

Figure G.5 displays the Klippel default windows for any individual measurement. You are at liberty to change the layout or the individual windows that you feel are more appropriate for your needs, however we would recommend that you save such changes as your own user templates. Having done so, you can use the next icon, shown as figure G.6 to save these for future use.

Figure G.6: Save Default Windows.

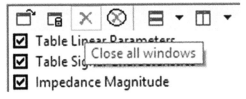

Figure G.7: Close All Windows.

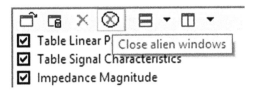

Figure G.8: Close Alien Windows.

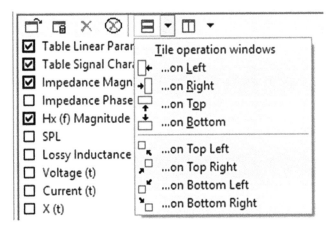

Figure G.9: Tile Horizontally.

The next icon closes *all* the measurement windows and is shown as figure G.7. Of more use if you are comparing lots of data is the next icon figure G.8; this enables you to close windows that are *not* related to the current measurement or window.

The next two icons allow you to tile the open windows horizontally or vertically. They are shown as figures G.9 and G.10.

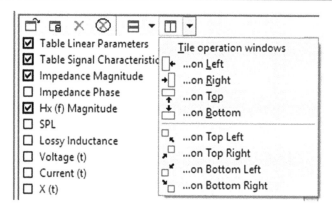

Figure G.10: Tile Vertically.

G.2   LPM—Example Measurement

Looking back at the Klippel LPM Thiele/Small Parameters shown in figure G.4 on page 268, we can see that although there is a lot of information, arguably too much, there is very little clarity in what is displayed. Partly this is inevitable, as each measurement contains both set-up and information about the measurement as well as the results of the measurement that engineers or customers are actually interested in.

Before we go any further, we need to know that any measurement has three essential components:

1.   Setting up the instrument.
2.   Information about the current measurement.
3.   Results of the measurement.

However, before we can start a measurement it is generally a good idea to collect some core information. Click on the properties icon; this is shown as figure G.11.

There are six tabs here that are used to configure the test and define information about the test object. The first of these is Info, shown as figure G.12. This is a text field where you can record what a measurement is about as well as when it was created, modified, or measured. We will describe in some detail each of these tabs for the LPM measurement, but please note that different information is recorded for other measurements and simulations.

The next tab is Driver, where we record physical information such as the diaphragm area S_d, diameter d_d, and material of the voice coil (copper or aluminium)—this is used to identify the voice coil temperature through the temperature coefficient—as well as the rated values power P_e and nominal impedance Z_n as shown in figure G.13.[2]

The Stimulus tab, shown in figure G.14, sets the conditions under which any given measurement is conducted: F max, frequency resolution, voltage, and averaging are just some typical options available here.

The Input tab, shown in figure G.15, selects the speaker routing and laser and/or microphone inputs.

Figure G.11: Klippel: Measurement Properties.

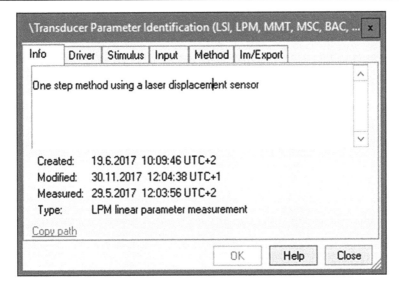

Figure G.12: Properties: Info.

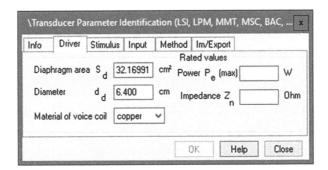

Figure G.13: Properties: Driver.

The Method tab, shown in figure G.16, selects between four methods of measuring the parameters:

1. Using laser.
2. Using additional mass.
3. Using a sealed enclosure.
4. Fixed Mmd.

As well as selecting between measurements in free air or in a sealed enclosure, there is the option to enter a series resistance to allow for lead resistance.

The next tab allows import and export of the settings and main measurement data along with using known values of $Bl(x = 0)$, Mms, or Re as well as using an imported displacement curve. It is shown in figure G.17.

Why would we do this? Well, in the case of a very small speaker it may not be possible to calculate the mass, but it may be easy to weigh it, or in the case of a loudspeaker with high inductance, we may not be able to accurately measure it dynamically. In any of these cases we can import a *known measured value* instead; similarly, we often export the LPM data into the LSI module when measuring the dynamic displacement characteristics, as the parameters are more accurately measured at low levels.

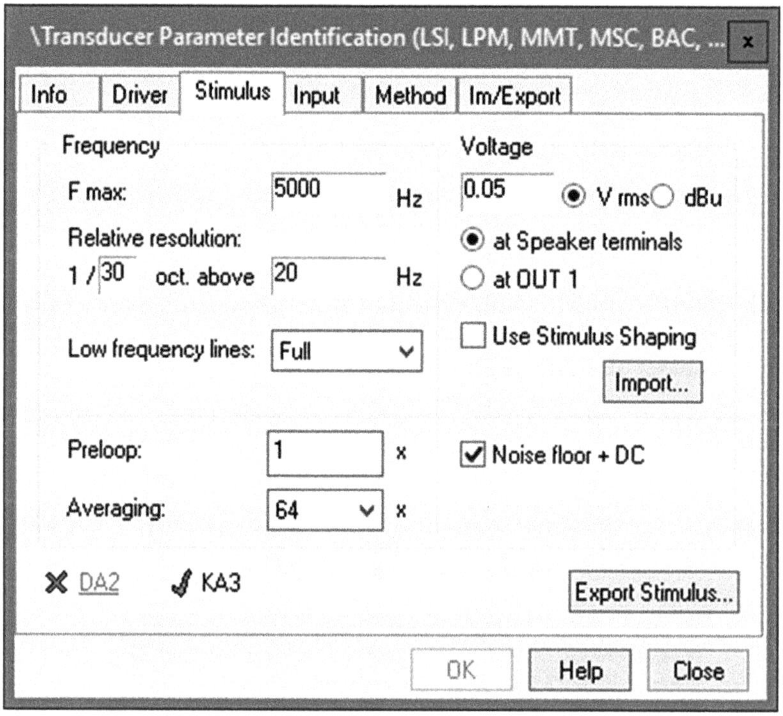

Figure G.14: Properties: Stimulus.

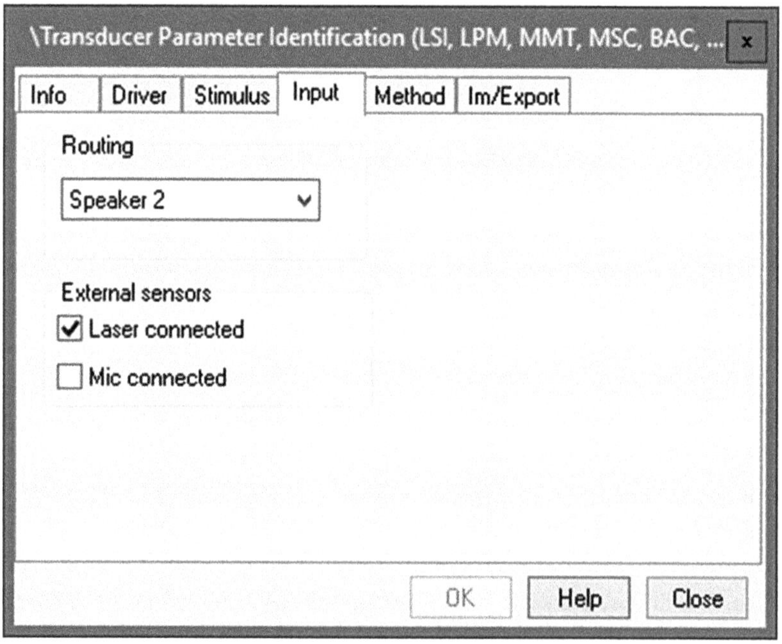

Figure G.15: Properties: Input.

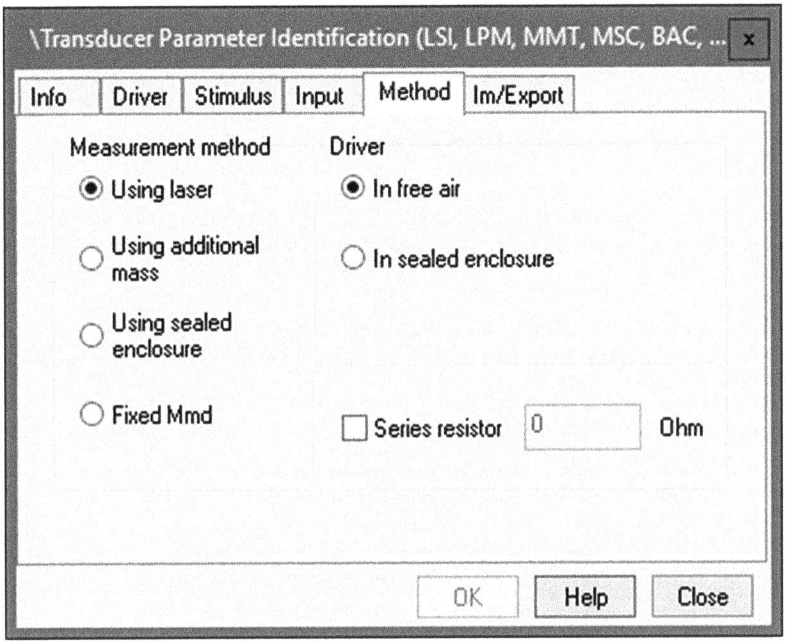

Figure G.16: Properties: Method.

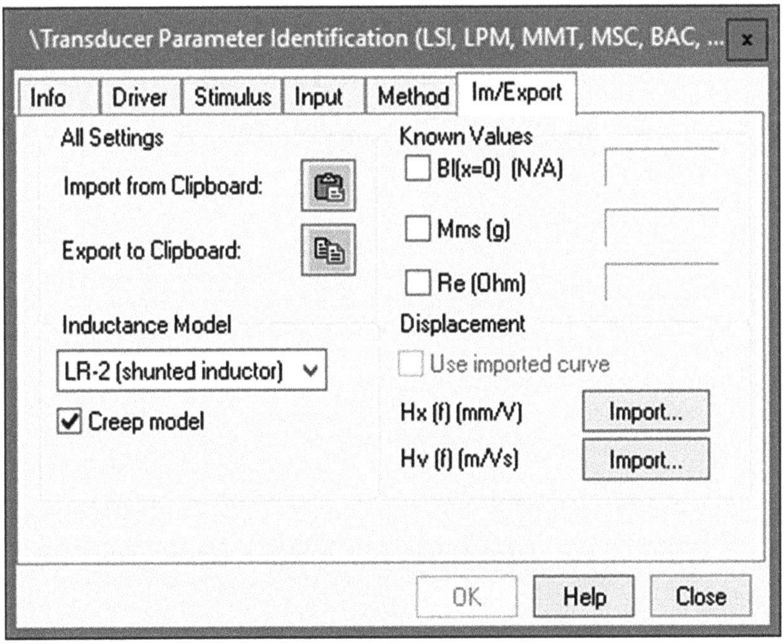

Figure G.17: Properties: Import/Export.

G.3   LPM Example Measurement Results

Before we run a measurement, let us examine the example database in detail. We will look first at the LPM results as shown in figure G.18. This is a combined table combining electrical, mechanical, loss factors, and other parameters in one place. Next is the parameter identification table; this is shown as figure G.19. This records set-up and information about the measurement.

The electrical impedance versus frequency is shown in figure G.20. This figure also shows the measured curve in black as well as the fitted curve from the underlying small signal model using the electrical parameters shown in figure G.19 in red.

Next is the transfer function Hx. This equals the displacement X(f) divided by the stimulus voltage V(f); this is shown in figure G.21. There are three curves here. The ragged black one is the measured laser displacement raw data, whilst

Name	Value	Unit	Comment
Electrical Parameters			
Re	4.86	Ohm	electrical voice coil resistance at DC
Le	0.156	mH	frequency independent part of voice coil inductance
L2	0.234	mH	para-inductance of voice coil
R2	2.96	Ohm	electrical resistance due to eddy current losses
Cmes	300.36	µF	electrical capacitance representing moving mass
Lces	8.10	mH	electrical inductance representing driver compliance
Res	17.24	Ohm	resistance due to mechanical losses
fs	102.1	Hz	driver resonance frequency
Mechanical Parameters			
(using laser)			
Mms	2.894	g	mechanical mass of driver diaphragm assembly including air load and voice coil
Mmd (Sd)	2.724	g	mechanical mass of voice coil and diaphragm without air load
Rms	0.559	kg/s	mechanical resistance of total-driver losses
Cms	0.840	mm/N	mechanical compliance of driver suspension
Kms	1.19	N/mm	mechanical stiffness of driver suspension
Bl	3.104	N/A	force factor (Bl product)
Lambda s	0.073		suspension creep factor
Loss factors			
Qtp	0.731		total Q-factor considering all losses
Qms	3.321		mechanical Q-factor of driver in free air considering Rms only
Qes	0.936		electrical Q-factor of driver in free air considering Re only
Qts	0.730		total Q-factor considering Re and Rms only
Other Parameters			
Vas	0.9507	l	equivalent air volume of suspension
n0	0.104	%	reference efficiency (2 pi-radiation using Re)
Lm	82.36	dB	characteristic sound pressure level (SPL at 1m for 1W @ Re)
Lnom	Zn missing	dB	nominal sensitivity (SPL at 1m for 1W @ Zn)
rmse Z	2.42	%	root-mean-square fitting error of driver impedance Z(f)
rmse Hx	2.65	%	root-mean-square fitting error of transfer function Hx (f)
Series resistor	0.00	Ohm	resistance of series resistor
Sd	28.27	cm²	diaphragm area

Figure G.18: LPM Results Table.

① Reduce Fmax to 20* fs to improve impedance fitting

Name	Value	Unit	Comment
Voltage			
U pp	0.38	V	peak to peak value of voltage at terminals
U ac	0.05	V rms	AC part of voltage signal
U dc	0.00	V	
U head	68.7	dB	digital headroom of voltage signal
U SNR+D	41.6	dB	ratio of signal to noise+distortion in voltage signal
fu noise	204.0	Hz	frequency of noise+distortion maximum in voltage signal
gain	18.80		Amplifier gain measured
Current			
I pp	0.06	A	peak to peak value of current at terminals
I ac	0.01	A rms	AC part of current signal
I dc	0.00	A	
I head	73.6	dB	digital headroom of current signal
I SNR+D	31.3	dB	ratio of signal to noise+distortion in current signal
fi noise	104.4	Hz	frequency of noise+distortion maximum in current signal
Displacement			
X pp	0.12	mm	peak to peak value of displacement signal
X ac	0.02	mm rms	AC part of displacement signal
X dc	0.00	mm	
X head	36.2	dB	digital headroom of displacement signal
X SNR+D	23.8	dB	ratio of signal to noise+distortion in displacement signal
fx cutoff	355.2	Hz	frequency of highest valid line in displacement signal
SPL			
p pp		mV	peak to peak value of microphone signal
p ac		mV rms	AC part of microphone signal
p head		dB	digital headroom of microphone signal
p sum level		dB	sum level of fundamentals in microphone signal
p mean level		dB	mean level of fundamentals in microphone signal
Measurement			
f sample	48000	Hz	sample frequency
N stim	131072	samples	stimulus length
cal x laser	0.000000		Laser calibration factor

Figure G.19: Parameter Identification.

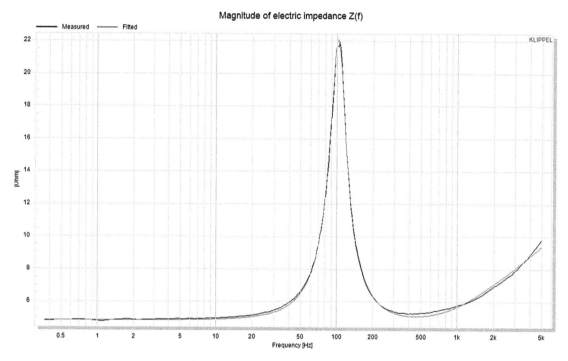

Figure G.20: Impedance Measurement.

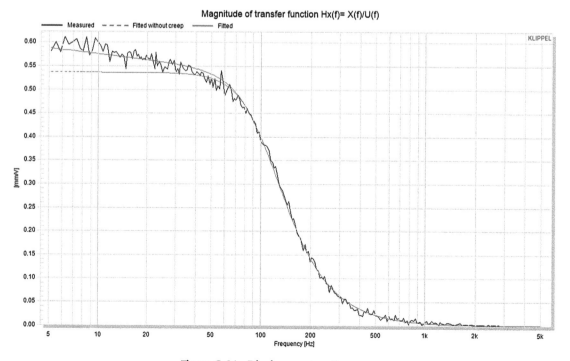

Figure G.21: Displacement vs Frequency.

the mauve curve is a fitted curve that is the mathematical model which most closely follows this data. Lastly there is a smooth black curve that is the modelled data without creep.

Figure G.22 shows the current spectrum at the speaker terminals in red, against the noise floor in black and the Noise+Distortion in grey. It is important that the Noise+Distortion is well below the signal level, preferably almost down to the noise floor, as shown in this example.

Lastly, in figure G.23 is shown the spectrum of the displacement vs frequency (Hz).

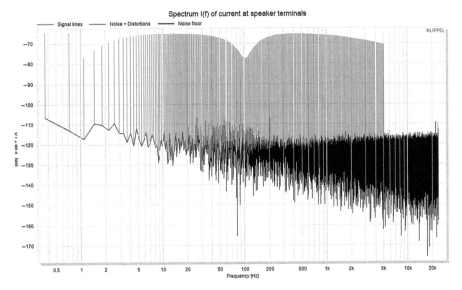

Figure G.22: Current vs Frequency.

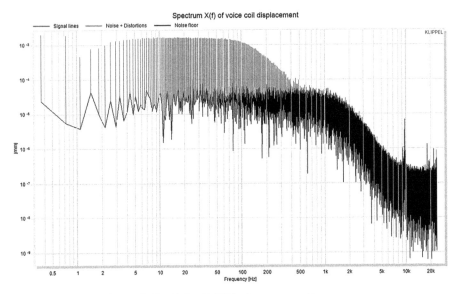

Figure G.23: Voice Coil Displacement vs Frequency.

There are other optional curves that can be selected, such as voltage (f) spectrum as shown in figure G.24. These can be examined to further investigate details if required.

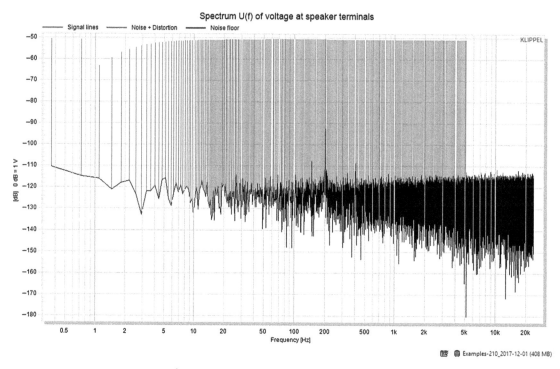

Figure G.24: Voltage Spectrum vs Frequency.

G.4 Running a LPM Measurement

The first thing we will do is to *duplicate* the measurement above. Left click on the measurement you wish to select, then *right* click and select Duplicate. This produces an identically set up test but without any measurement results. This is shown in figure G.25.

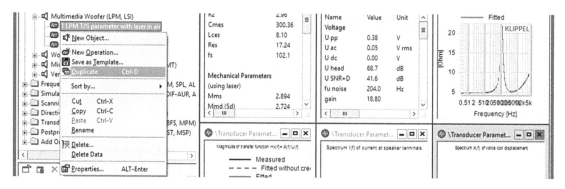

Figure G.25: Duplicate LPM Test.

The resultant *blank* test is shown in figure G.26.

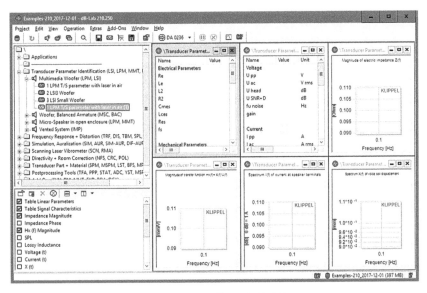

Figure G.26: Duplicated LPM Test.

G.5 Small Signal Measurements on a SEAS H1207

We now run the test on a SEAS H1207 drive unit by clicking on the Run icon.[3] The result is shown in figure G.27.

Notice we have one warning in pale yellow above the Parameter Identification window asking you to reduce Fmax to less than 20 * fs. This is purely to increase the match of the model to measured impedance above 1 kHz. In this case we can safely ignore this warning. The final measurement is shown in figure G.28.

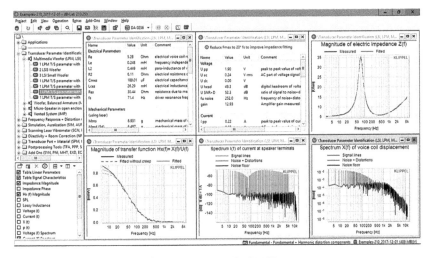

Figure G.27: SEAS LPM Test.

Name	Value	Unit	Comment
Electrical Parameters			
Re	5.28	Ohm	electrical voice coil resistance at DC
Le	0.248	mH	frequency independent part of voice coil inductance
L2	0.449	mH	para-inductance of voice coil
R2	6.11	Ohm	electrical resistance due to eddy current losses
Cmes	189.01	µF	electrical capacitance representing moving mass
Lces	26.29	mH	electrical inductance representing driver compliance
Res	30.44	Ohm	resistance due to mechanical losses
fs	71.4	Hz	driver resonance frequency
Mechanical Parameters			
(using laser)			
Mms	6.931	g	mechanical mass of driver diaphragm assembly including air load and voice coil
Mmd (Sd)	6.497	g	mechanical mass of voice coil and diaphragm without air load
Rms	1.205	kg/s	mechanical resistance of total-driver losses
Cms	0.717	mm/N	mechanical compliance of driver suspension
Kms	1.39	N/mm	mechanical stiffness of driver suspension
Bl	6.055	N/A	force factor (Bl product)
Lambda s	0.064		suspension creep factor
Loss factors			
Qtp	0.382		total Q-factor considering all losses
Qms	2.581		mechanical Q-factor of driver in free air considering Rms only
Qes	0.447		electrical Q-factor of driver in free air considering Re only
Qts	0.381		total Q-factor considering Re and Rms only
Other Parameters			
Vas	2.8296	l	equivalent air volume of suspension
n0	0.221	%	reference efficiency (2 pi-radiation using Re)
Lm	85.65	dB	characteristic sound pressure level (SPL at 1m for 1W @ Re)
Lnom	87.46	dB	nominal sensitivity (SPL at 1m for 1W @ Zn)
rmse Z	4.64	%	root-mean-square fitting error of driver impedance Z(f)
rmse Hx	3.39	%	root-mean-square fitting error of transfer function Hx (f)
Series resistor	0.00	Ohm	resistance of series resistor
Sd	52.81	cm^2	diaphragm area

Figure G.28: SEAS H1207 LPM Measurement.

SEAS PRESTIGE

L12RCY/P
H1207

Stiff, yet light aluminum cone and low loss rubber surround show no sign of the familiar 500-1500 Hz cone edge resonance and distortion associated with soft cones.

High temperature voice coil wound on an aluminuim voice coil former gives high power handling capacity.

Bullet shaped phase plug reduces compression due to temperature variations in the voice coil, avoids resonance problems which would occur in the volume between the dust cap and the pole piece and increases the long term power handling capacity.

Extra large magnet provides high efficiency and low Q.

This unit may be used in very small two-way ported systems producing an astonishingly deep bass and a clean, neutral midrange.

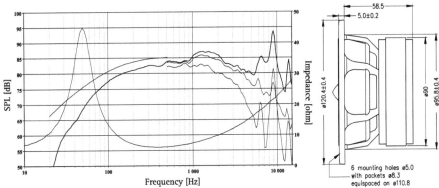

Frequency [Hz]

The frequency responses above show measured free field sound pressure in 0, 30, and 60 degrees angle using a 2.5L closed box. Input 2.83 VRMS, microphone distance 0.5m, normalized to SPL 1m. The dotted line is a calculated response in infinite baffle based on the parameters given for this specific driver. The impedance is measured in free air without baffle using a 2V sine signal.

Nominal Impedance	8 Ohms	Voice Coil Resistance	5.5 Ohms
Recommended Frequency Range	55 - 3500 Hz	Voice Coil Inductance	0.76 mH
Short Term Power Handling *	200 W	Force Factor	6.1 N/A
Long Term Power Handling *	70 W	Free Air Resonance	50 Hz
Characteristic Sensitivity (2,83V, 1m)	85.5 dB	Moving Mass	6.8 g
Voice Coil Diameter	26 mm	Air Load Mass In IEC Baffle	0.21 g
Voice Coil Height	12 mm	Suspension Compliance	1.5 mm/N
Air Gap Height	6 mm	Suspension Mechanical Resistance	0.94 Ns/m
Linear Coil Travel (p-p)	6 mm	Effective Piston Area	50 cm²
Maximum Coil Travel (p-p)	9 mm	VAS	5 Litres
Magnetic Gap Flux Density	1.1 T	QMS	2.34
Magnet Weight	0.42 kg	QES	0.33
Total Weight	1.21 kg	QTS	0.29

Jul 2007-1   *IEC 268-5   W12-411
SEAS reserves the right to change technical data

RoHS compliant product

www.seas.no

Figure G.29: SEAS Datasheet.

This is very close to the SEAS data sheet values as show, in figure G.29. Notice however that the drive unit resonance is higher than the data sheet, indicating that this driver has not been used hard or run in!

Notes

1. The Klippel LPM module goes well beyond the Thiele/Small parameter set, which is best thought of as a subset of loudspeaker parameters.
2. The diaphragm area and diameter are linked text inputs and can be used to calculate the other variable given a round loudspeaker.
3. Here we are using DA0236.

Mecway Tutorial

H.1   Introduction

In this chapter we use Mecway to build a model of our subwoofer driver: entering nodes and elements then assigning materials, loads, and constraints. Mecway is available free as a 1000-node version. Full 32-bit or 64-bit versions are available from www.mecway.com/.

Ultimately our aim is to develop a dynamic model which enables us to visually see the physical behaviour of a loudspeaker drive unit or parts of one. Such an analysis brings real understanding to the process of design.

Mecway has a wide range of expanding capabilities including:

- Static
- Thermal
- Modal Vibrational
- Dynamic Response
- Nonlinear
- Fluid
- Buckling
- Electric
- Magnetic
- Acoustic

Although sound and therefore acoustics is the aim of this book, we feel that any finite element analysis (FEA) program is not best suited to acoustics due to the normal tendency of such systems to be 'open' or without defined boundaries. Therefore, the acoustics will be tackled by a boundary element modeller (BEM).

We will however in this tutorial show modelling the mechanics, the static, modal, dynamic, buckling, and nonlinear models. We are aiming to export the vibration modes and positions for later processing with other tools.

In this tutorial, the models have been made using Mecway version 4. It shows the steps we have used in producing working models.

H.2   MecWay Graphical User Interface

There is a standard Windows interface with just five menu choices, littered with lots of unusual icons. First, let's explore what we have got and what we can find out about the interface.

H.2.1   Menu Structure

The File menu structure is shown as figure H.1.

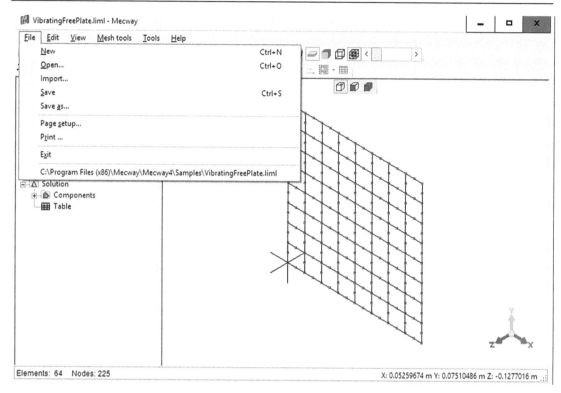

Figure H.1: Mecway: File.

Nothing unusual here except maybe 'Load into model'.

Program Files\Mecway\Mecway4\Samples is where to look for examples of working models . . .

The Edit menu structure is shown as figure H.2 and the View menu structure is shown as figure H.3.

Again pretty straightforward—but what is 'Open cracks'? Let us hope it is to show minor gaps in between nodes in a mesh or something like that—as that has been a regular pain of using FEA for years!

The 'Mesh tools' menu structure is shown as figure H.4.

Note the Mesh tools menu becomes 'greyed' out when a solution is present. There are lots of choices here, obviously to be used purely at the model and meshing stage.

The Tools menu structure is shown as figure H.5.

There do not appear to be many here and no obvious guide as to how to use them.

H.3   Icons

Top Line: New ⬜ , Open 📂 , Save 💾 , Undo ↶, Redo ↷ are as you would expect, then we get to the more interesting ones: Solve ☰, Isometric View ⬡ , XY View ⬡ , Fit to screen ⧉ , Zoom 🔍 , Rotate ⟳ ,

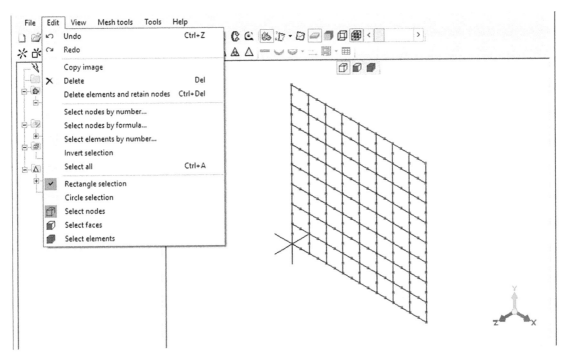

Figure H.2: Mecway: Edit.

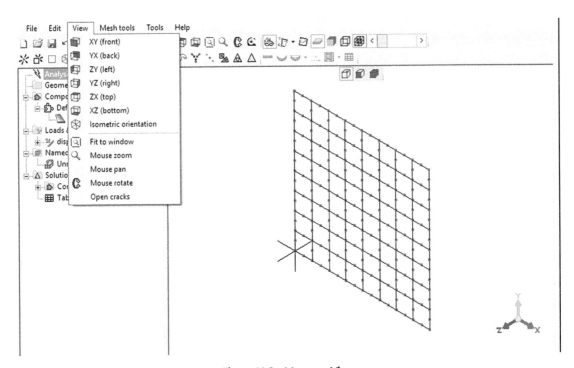

Figure H.3: Mecway: View.

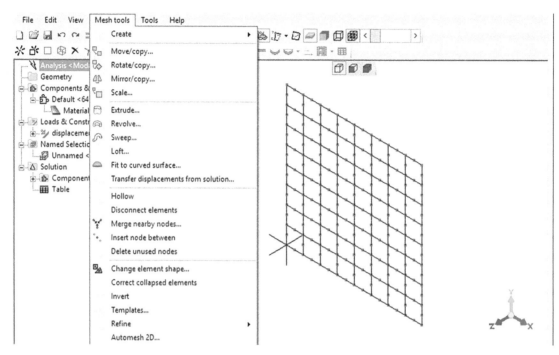

Figure H.4: Mecway: Mesh Tools.

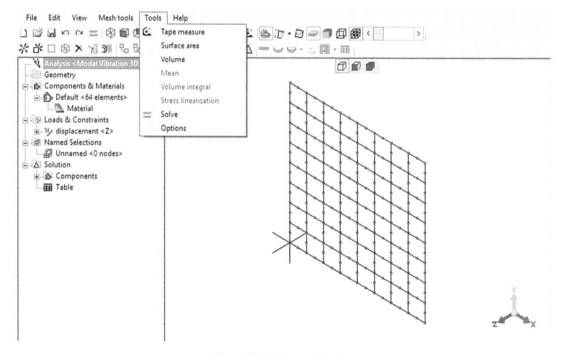

Figure H.5: Mecway: Tools.

Tape measure , Show loads and constraints , Show node and element numbers , Show element axes , Show shell thickness , Show element surfaces , Show element edges , Cutting plane

Next Line Down: New node , New element , Polyline , Quick square , Quick cube , Delete nodes/elements , Node coordinates , Element properties , Move/copy , Rotate/copy , Extrude , Revolve , Merge nodes , Insert node between , Change element shape , Refine Element .

The next icons are blank at first and only come 'alive' with a solved model. View undeformed , View deformed , View undeformed mesh on or off , Show arrows[1] , and finally Animation[2] .

H.4 Menu Tree

H.4.1 Analysis

Click on File->New. This gives us a blank model as shown in figure H.6.

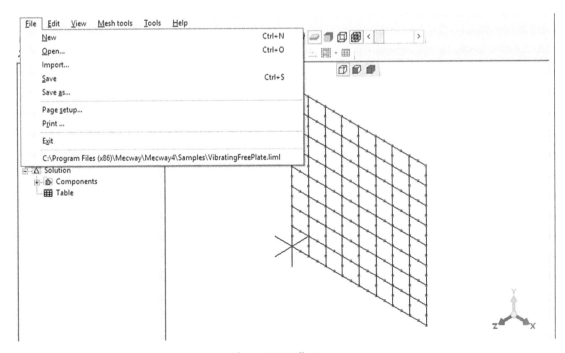

Figure H.6: File New.

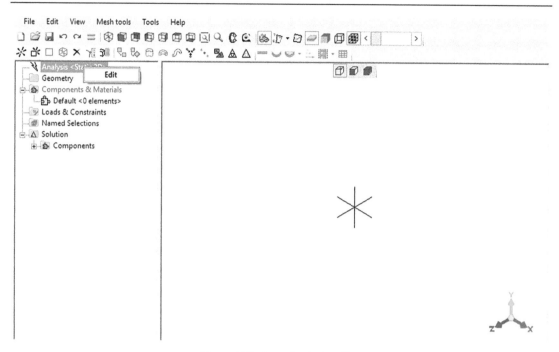

Figure H.7: New Analysis Edit.

Initially we get a blank Analysis <Static 3D>, so right click over the Analysis <Static 3D> and Edit appears, as shown in figure H.7.

Click on Edit and Global Properties appears, as shown in figure H.8.

3D comes up by default. We could choose 2D or 3D, however, for the moment, we're going to go with Axisymmetric->Dynamic Response Axisymmetric, as shown in figure H.9.

Then click OK and we go back to the graphical user interface as shown in figure H.10.

We can see an extra box has appeared below the main screen—don't worry about this at the moment.

H.4.2   Geometry

If you right click on Geometry, you get the chance to Import STEP/IGES file. We have not got either available, just dwg and dxf. But this is where MecWay's later versions will help us, as MecWay will directly open *.dxf files.

If you have a suitable dxf file, just select File->Import and navigate to your file. This will save us the need to use a STEP/IGES compatible CAD program like FreeCAD.

So we will not need to do it the old-fashioned way and construct it from scratch. Although, as we have said in earlier chapters, this would not necessarily be a bad thing as it would force you to make a simple model. This will generally solve more quickly and will usually be simpler to debug.

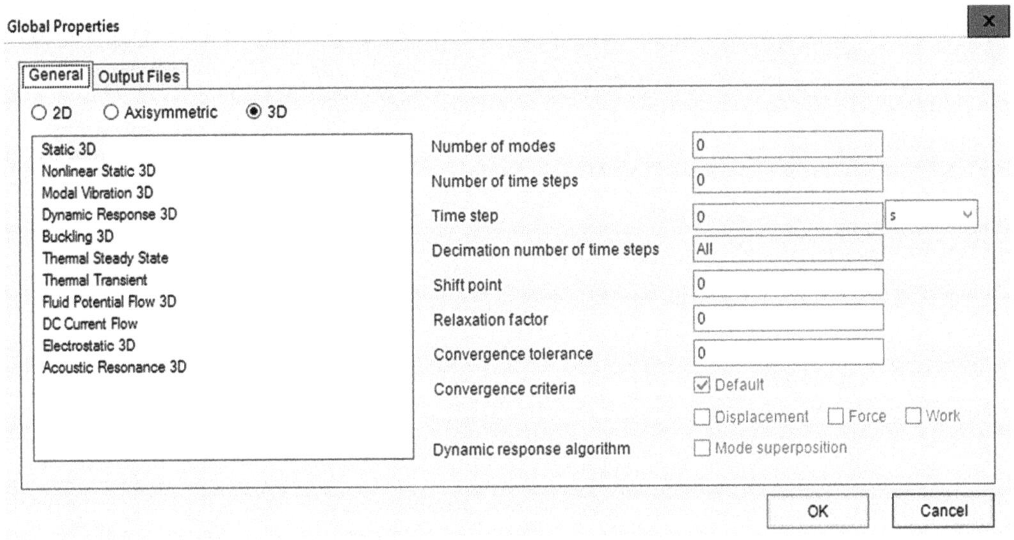

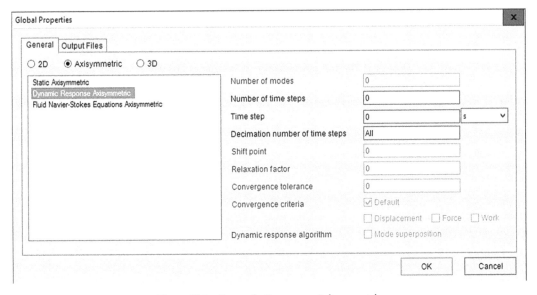

Figure H.8: Global Properties.

Figure H.9: Dynamic Response Axisymmetric.

H.4.3   Components & Materials

This is where we will define the component parts in the model and then the materials that will comprise it.

You need to do it this way at first even though often the same material, maybe with different thicknesses, will be used for different parts in even a simple model.

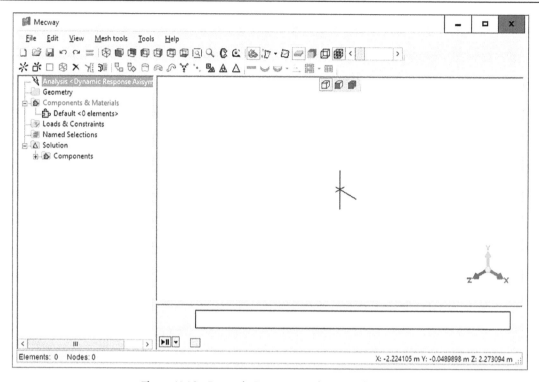

Figure H.10: Dynamic Response Axisymmetric GUI.

Notice that at first, Components & Materials is in red. This is the program's way of saying 'Hey! There is something wrong here!'

Right click on Components & Materials and then click 'New component'. Do this for as many types of components as your model will contain, as shown in figure H.11.

Right click the new component and after ensuring that the visible check box is *ticked*, right click 'Assign new material'. This then displays the Material Properties (Material) dialogue shown in figure H.12.

It is at this point that we need to start entering some information about our materials. There are five tabs:

1.   Geometric
2.   Mechanical
3.   Thermal
4.   Fluid
5.   Electromagnetic

H.4.4   Loads & Constraints

Right click on Loads & Constraints for the choices shown in figure H.13.

H.4.5   Named Selections

Right click on Named Selections for the choices shown in figure H.14.

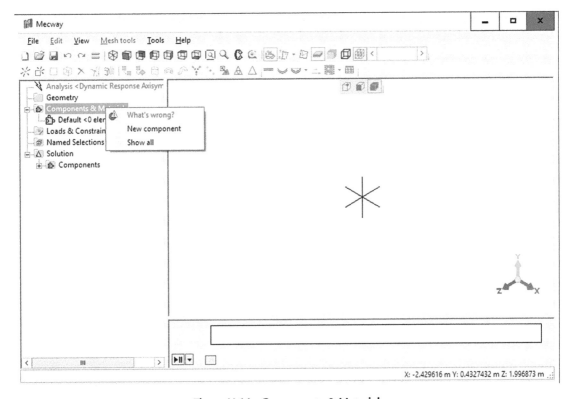

Figure H.11: Components & Material.

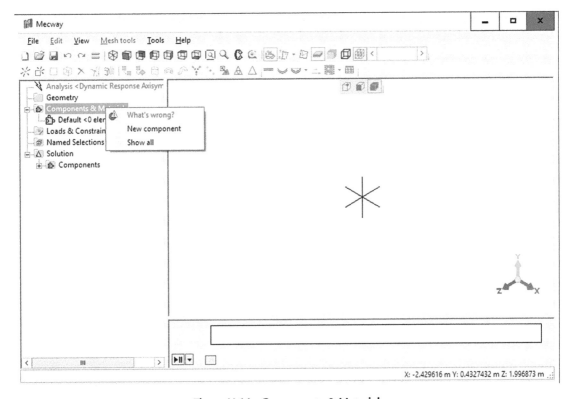

Figure H.12: Material Properties Dialogue.

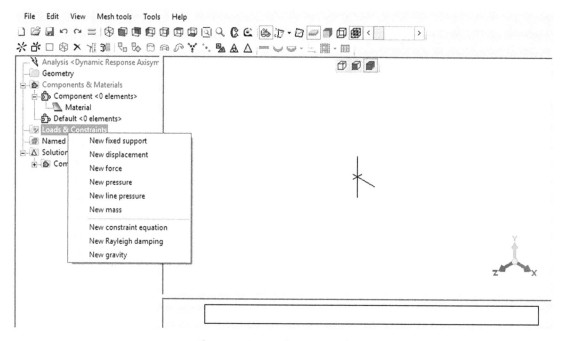

Figure H.13: Loads & Constraints.

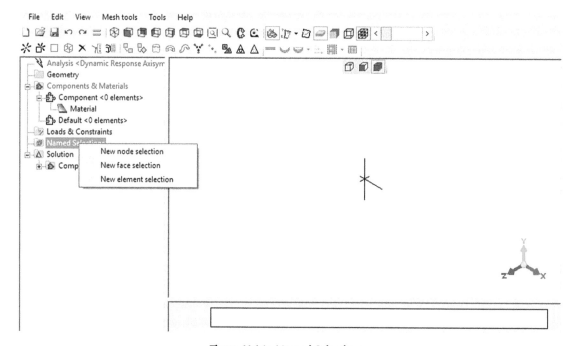

Figure H.14: Named Selections.

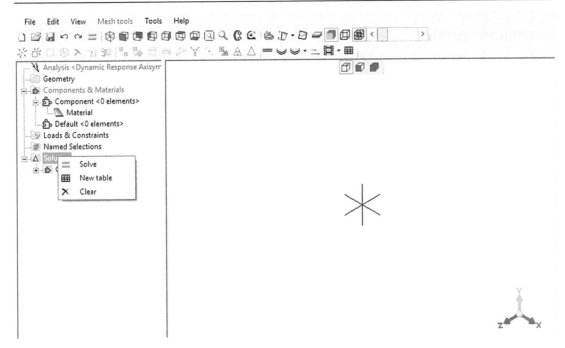

Figure H.15: Solution Solve.

H.4.6 Solution

Lastly, right click on Solution for the choices shown in figure H.15.

We will cover all of these as we build our model.

H.5 Subwoofer Driver Model

So let us start with the end in mind—what we are going to do is model the mechanical vibrations of a loudspeaker voice coil and former, cone, and surround. For the moment, we will use Mecway's node entry system to create a very basic voice coil, former, cone, and surround axisymmetrical model—ignoring all of the other parts. It's a bit of a cheat but it should enable us to model the essential components to start with at first.

Also note that this model will use *perfect* coupling between the parts. It is here that we depart from some colleagues in the industry whose aim seems to be to include lots of detail. For example, glue joints in order to make the simulation as accurate as possible; whilst this is a laudable aim in theory, we feel it is more important to ensure the underlying performance is modelled efficiently, even if it is not perfect. Yes, you can build extremely accurate models that represent everything—but the time involved can be excessive if you are not careful.

We believe our aim should be first to get a model that works reasonably accurately. We can develop and refine it later on.

We will use an axisymmetric model. The voice coil is copper, the former and cone are both aluminium, whilst the surround is rubber.

H.5.1   Materials in Model

The material parameters we will use are shown in table H.1.

Table H.1: Material Parameters.

Material	Density kg/m^3	Young's Modulus GPa (e.g. 100e9)	Poisson's Ratio	Damping
Aluminium	2700	69e9	0.33	0.0001
Copper	8700	110e9	0.33	0.002
Rubber	1100	0.01e9–0.1e9	0.48 *(NOT 0.5!)*	0.01–0.08
Spider	1400	0.43e9	0.33	0.1

H.5.2   Initial Nodes

We will first start off with nodes at the positions shown in table H.2 and continue with table H.3 on page 295.

To enter these: Mesh Tools->Create->Nodes. You need to enter the X, Y, and Z coordinates if you are doing it this way.

Let's set up a model now. We have some coordinates above; we will enter them using 'New node' ✳. Ideally, we might want to produce a 3D model but we will start off with an axisymmetric one and revolve it into 3D. It should look like figure H.16.

Often the view first displayed is an Isometric. Click View->XY (front) view and then View->Fit to window. It should then look like figure H.17.

Table H.2: Subwoofer Driver Voice Coil and Cone Nodes.

Node#	Element#	Part	Name	X	Y	Z
0		Origin		0	0	0
1		Y-Axis		0	0.1	0
2		Voice Coil		0.0375	−0.016	0
3		Voice Coil		0.0375	0.016	0
4		Voice Coil		0.03875	−0.016	0
5		Voice Coil		0.03875	0.016	0
6		Former	Top Former	0.0375	0.036	0
7		Former	Top Former	0.0374	0.036	0
8		Former	Bottom Former	0.0374	−0.016	0
9		Former	Bottom Former	0.0375	−0.016	0
10		Cone	Bottom Neck	0.0375	0.030	0
11		Cone	Bottom Neck	0.0377	0.030	0
12		Cone	Top Neck	0.0375	0.0362	0
13		Cone	Top Neck	0.0377	0.0361	0
14		Cone	Top Neck	0.0377	0.036	0
15		Cone	Top	0.1588	0.1086	0
16		Cone	Top	0.1588	0.1083	0

Table H.3: Subwoofer Driver Surround Nodes.

Node#	Element#	Part	Name	X	Y	Z
17		Surround R8	1st	0.15800	0.10961	0
18		Surround R8	2nd	0.15827	0.11068	0
19		Surround R8	3rd	0.15907	0.11361	0
20		Surround R8	4th	0.16034	0.11527	0
21		Surround R8	5th	0.16200	0.11654	0
22		Surround R8	6th	0.16393	0.11734	0
23		Surround R8	7th	0.16600	0.11761	0
24		Surround R8	8th	0.16807	0.11734	0
25		Surround R8	9th	0.17000	0.11654	0
26		Surround R8	10th	0.17166	0.11527	0
27		Surround R8	11th	0.17293	0.11361	0
28		Surround R8	12th	0.17373	0.11168	0
29		Surround R8	13th	0.17400	0.10961	0
30		Surround R8	14th	0.17400	0.10861	0
31		Surround R7	1st	0.17300	0.10861	0
32		Surround R7	2nd	0.17276	0.11068	0
33		Surround R7	3rd	0.17206	0.11261	0
34		Surround R7	4th	0.17095	0.11427	0
35		Surround R7	5th	0.16950	0.11554	0
36		Surround R7	6th	0.16778	0.11634	0
37		Surround R7	7th	0.16600	0.11661	0
38		Surround R7	8th	0.16419	0.11634	0
39		Surround R7	9th	0.16250	0.11554	0
40		Surround R7	10th	0.16105	0.11427	0
41		Surround R7	11th	0.15994	0.11261	0
42		Surround R7	12th	0.15924	0.11068	0
43		Surround R7	13th	0.15900	0.10861	0
44		Surr R7 - R8	14th	0.15800	0.10961	0

Next we need our three materials, (aluminium, copper, and rubber) and four, no let's make that five, components (former, voice coil, cone neck, cone body, and surround). We know the material parameters from the Material Parameters table H.1.

Notice both Analysis and Components & Materials are now red. Again, right clicking shows the 'What's wrong' dialogue. Just ignore this for the moment, as we will now construct some elements on our nodes.

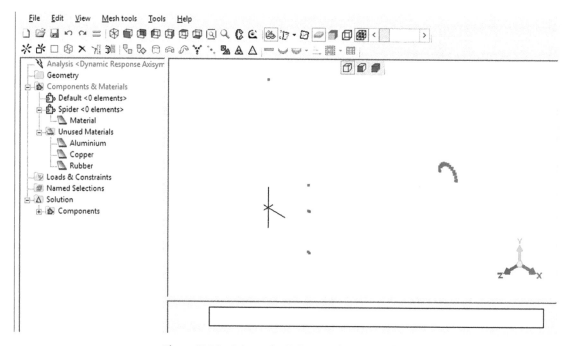

Figure H.16: Subwoofer Driver Nodes Isometric.

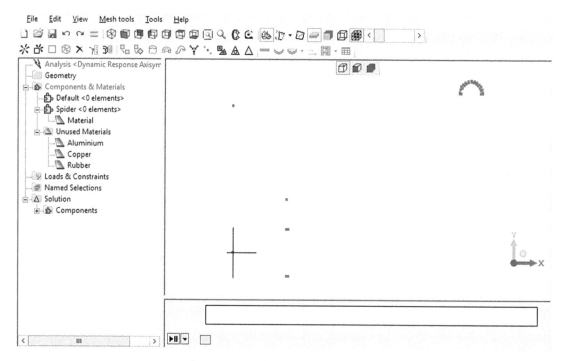

Figure H.17: Subwoofer Driver Nodes.

Click on Mesh tools->Create->Element.

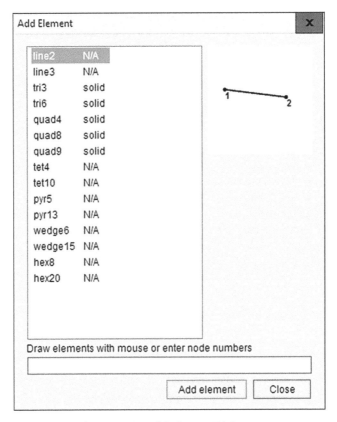

Figure H.18: Add Element Dialogue.

Click on 'line2' and then on Node0 (0, 0, 0) and then Node1 (0, 0.1, 0). A line appears between the origin and a point 0.1 above it, as shown in figure H.18.[3]

If you hold a mouse cursor over the area nodes of the voice coil you can zoom in using a mouse wheel to focus just on these nodes.

Click on Mesh tools->Create->Element then click on quad4 (this is a simple element to start with), though we can and will change it later. Starting at the bottom of the voice coil, join up the four nodes that make it up, as shown on the Add Element Dialogue. Click on New Element and a new dialogue will appear as in figure H.19.

Notice that as you click each node, they are added to the list at the bottom of the dialogue. We now continue to connect nodes as above to form them up into elements until we have figure H.20.

This gives us an Axisymmetric view of our proposed model.

Are we ready to go on yet? No, not yet—Mecway has one very good trick up its sleeve that is very useful. Go to View->Open cracks, as shown in figure H.21.

The programmers are being very good to us here by showing us where problems lie in this model before we go any further. Oh, the hours, or probably days, that this technique could have saved using other FEA programs!

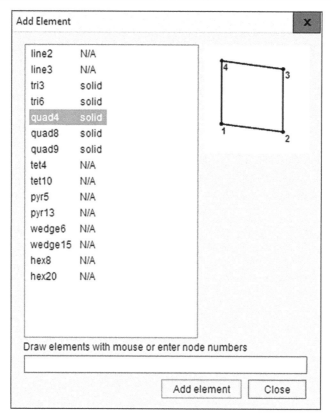

Figure H.19: Add Quad4 Element Dialogue.

Now let us go back to the original view: Can you see the gaps? Probably not—here we have computers and programs working literally as we have asked them to. The trouble is that some very small dimensions did not *exactly* match up.

Fortunately, Mecway is very good and figuratively says, 'Hey, the problem is somewhere here', and shows the affected areas in shades of grey or as elements narrowing to a point.

There are three areas these programs cannot handle unaided here:[4]

- The junction between the top edge of the cone and the surround.
- The junction between the top of the cone neck and the cone body.
- A gap between the top of the voice coil and the former.

Okay, I'm assuming that you have followed the detail (tedious though it is) so far. Let's proceed and fix our model.

We will start at the gap between the top of the voice coil and the former. The reason is, this gap is quite simple when we draw two very large elements, the former and the voice coil. We only actually joined them at one place, nodes 2 and 9, so the *top of the voice coil* is not actually connected to the *top of the former*. The best way to do this is to introduce some nodes at the top of the former, thus splitting the former into three sections.

So, let's create these new nodes ⁕. These are shown in tables H.4 and H.5 on page 300.

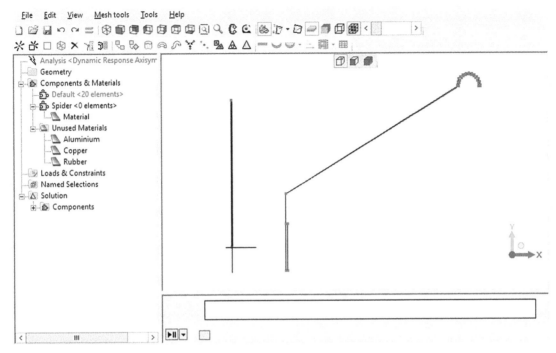

Figure H.20: Subwoofer Initial Elements.

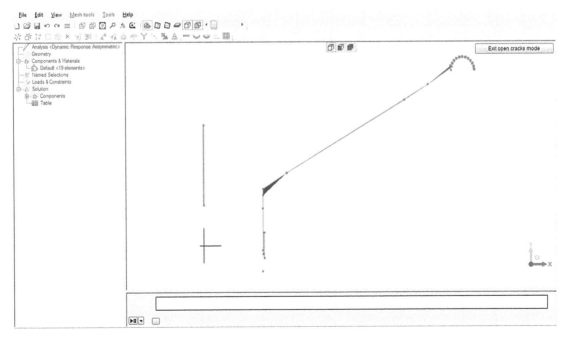

Figure H.21: Subwoofer Initial Elements Open Cracks.

Table H.4: Mecway Additional Nodes 1.

Node#	Element#	Part	Name	X	Y	Z
45		Former	Top Voice Coil	0.0374	0.016	0

Table H.5: Additional Nodes 2.

Node#	Element#	Part	Name	X	Y	Z
46		Former	Bottom of Neck	0.0374	0.03	0

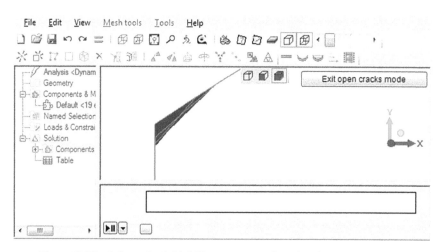

Figure H.22: Subwoofer Initial Elements Top of Neck using Open Cracks.

Now, click on 'Select elements' [icon]. All the nodes disappear but we can select the *elements*—zoom in and select the former and then delete it.[5] Click 'Select nodes' again [icon]. We can now rebuild the bottom section of the former with one element.

Then, looking upward toward the former to the cone neck joint, we can see one node all by itself. So let's make a matching node at the bottom of the neck in the same way as we did for the former and voice coil.

The next step is to rebuild the top and middle sections of the former with one element each.[6] Click 'Open cracks' again, it should look like figure H.22.

We will now need to refine the elements.

Starting with the former, so we will set the other components to invisible.

Select Mesh tools->Refine->Custom.

We are using quad4 elements so we need to use R and S. (We're not sure what these mean, though.)

Right click Components & Materials and then click 'New component'. This creates a new component which has no nodes or elements yet; we will assign some materials and properties for later use.

Click 'Assign new material', then click 'Plate/shell/membrane', as we will be using very thin materials for the former and the cone at least.

Click the Mechanical tab. Isotropic, Young's Modulus, Poisson's ratio, and Density are all available, so let's enter these for our three materials.[7]

For the moment, we will continue without the damping but this needs to be included later.

As there is no other information, we are going to assume that normal SI units are to be used: Young's modulus in GPa and density in kg/m³. Having entered the information for aluminium, click Add.

Notice also that the Thickness input box is now available. This is slightly problematic as the same material is often used in different section(s). Also, if the model has been correctly drawn or entered, this information is presented by the nodes. Else, if it's a known thickness, we need to define materials for each individual part.

That's what we are going to try—so four materials: aluminium former 0.0001m thick, copper, rubber 0.0005m thick, and aluminium cone 0.0005m thick.

Then enter the information for copper; this is for the voice coil wire and as it is not a Plate/shell/membrane, we were not sure how to enter it correctly.

Eventually, after much backward, forward, and right clicking, we have created all four major components and three materials.

We tried R = 10 and S = 2 but that divided the width by 10 and the length by 2, which makes no sense. So we then tried R = 2 and S = 10, which we can see if we look at the Nodes by clicking 'Select nodes'.

Save the model at this stage.

When we looked at the model though, *all* of the elements had been 'refined' in *exactly* the *same* way!

This is fine for the former, the voice coil, and the bottom of the cone. however, it is useless for the cone itself.

So we need to split the cone into separate sections: ConeNeck.

Then Refine Mesh Custom.

This is what I used to sub-divide the cone.

Make the former invisible and the voice coil visible and repeat the process.

We then revolved this model into a <Dynamic Response 3D> model below:

Ran it using sine wave steps from 0 degrees to 0/360 degrees and the model failed at the last step.

Well it looked okay but the Solver threw it out. . .

View>Open cracks - Shows that although the model has revolved 360 degrees, it has not actually joined up correctly.

Eventually we went back to our original un-revolved drawing, set that to static and tried to solve that with this result.

We then did an 'Open cracks' on this model, which is much simpler than the full 3D.

At first sight, it looks fine; however, it does seem to narrow down toward the neck so we zoom into the first element of the cone and it looks to be twisted, so it looks likely that this is where the problem lies.

Selecting the first element—we will try deleting it.

Now running 'Open cracks' again, we can see that the neck and former do not meet correctly either.

Eventually we tried putting extra nodes up the top of the former, matching those in the neck, but at definite distance and giving the top node a slight move inward to 'balance' the element shape by ensuring that only four nodes make up the element rather than five before.

'Open cracks' now looks more even.

So let's look below the neck. Here the problem seems to be that the nodes for the voice coil and the former do not match horizontally. So we delete and rebuild them so they are all parallel to each other.

There is still some slight distortion but no gaps showing.

Trying to Solve it but. . .

But at least we can see which nodes are causing problems: 130, 129, and 130.

Clicking Correct collapsed elements. . .

After Cleaning up and reassigning the new or repaired elements to the correct component.

It still does not run but there are no error messages. . .

Changing the element shape from quad4 to quad8, the model ran but gave no real outputs.

Axisymmetric Models that work. . . #12 is the first one that really looks correct. . .

The key was setting the Global Properties correctly.

Then the Applied force 'AppliedSine' though it is actually a Triangle Waveform.

3D Models Following on from the successful models above, we tried again revolving this into a full 360 degree 3D model, though it looks okay a gap remains at 0/360 degrees which has defeated us at the moment.

However, we can also apply cyclic symmetry, whereby a section is modelled but not the whole thing. So in theory it should be possible to generate a small section 15 to 45 degrees. We started with 15 degrees using a 3D vibration solid & truss.

Applying constraints and loads as shown, we have not yet tried the cyclic symmetry.

However, we can clearly see 'Modes' in the Surround e.g. 10th at 886 Hz.

Cyclic Symmetry does not appear to be available in Version 8. However, Mirror Symmetry is, so instead we can model a 90° section. This is shown as figure H.23.

Once we have our model running correctly, we can then start to analyse this case. We will look at the cone and surround physically moving. This is shown as figure H.24.

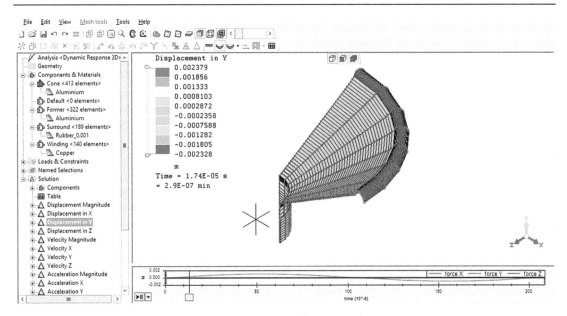

Figure H.23: Subwoofer 90° Section.

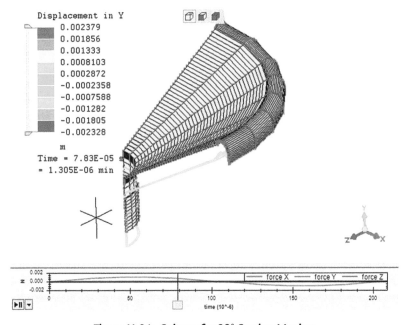

Figure H.24: Subwoofer 90° Section Moving.

Notes

1. When vector magnitude is selected.
2. Needs Table clicking before it activates.
3. We will use this line later as our line of axis symmetry for rotation.

4. If you still cannot see them then please zoom into these areas until you do.
5. Please ensure you are in Element Mode at this point.
6. Yes, we know that we will need more elements in the model, but we would like show the problems en route.
7. Damping coefficient is not available but Rayleigh damping *is* available in Mecway.

Micro-Cap Tutorial

Micro-Cap is a Simulation Program with Integrated Circuit Emphasis (SPICE) simulator or, more accurately, a lumped parameter modelling tool which we will use for relatively simple electrical crossovers and loudspeaker models. The full version is capable of much, much more, but that is outside the scope of this book.

It is available as a free download from Spectrum Software at www.spectrum-soft.com/index.shtm.

Micro-Cap is currently at version 11. The techniques developed in this book have been used successfully from version 6 onwards.

Micro-Cap is designed for simulation of electronic circuits, both analog and digital, with an emphasis on integrated circuit design where tens of thousands of transistors or gates are regularly simulated. A demo version limited to 50 components and a few other restrictions may be freely available for download for anyone to use for any purpose.

Fifty components may not appear to be much use, but with a little thought it is plenty for modelling quite complex loudspeakers and crossovers, and also electronic filters and amplifiers.

Micro-Cap uses a drag-and-drop editor where you can select the Component>Analog Primitives>Passive Components:

- Resistor
- Capacitor
- Inductor

The Micro-Cap 11 GUI is shown as figure I.1 on page 306 to show how to navigate the menus as the passive components are called.

As our goal is to produce a loudspeaker, as part of that it will be essential to predict the overall performance as a system. To do this we will probably need to integrate the separate loudspeaker drivers together using a crossover.

Why is it called a crossover? Quite simply, it is a frequency point whereupon one loudspeaker driver's output begins to cease and the output of another driver, often with different characteristics, starts to increase.

When passive components—resistors, capacitors, and inductors—are inserted between an amplifier and the loudspeaker(s), we then have a *passive* crossover.

When these passive components are instead replaced with active circuits[1] these are placed before the amplifier(s) and we then then have an *active* crossover.

Why these two methods? To understand this, we need to look back in time to the dawn of electronics and of loudspeakers themselves, back in fact to the 1920s. Back then, amplifiers were rare and expensive.

It's an often overlooked fact that many of the first loudspeakers were actually active systems. Amplifiers back then were both expensive and low powered, so the passive crossover was designed to split the signal passively to feed multiple loudspeakers.

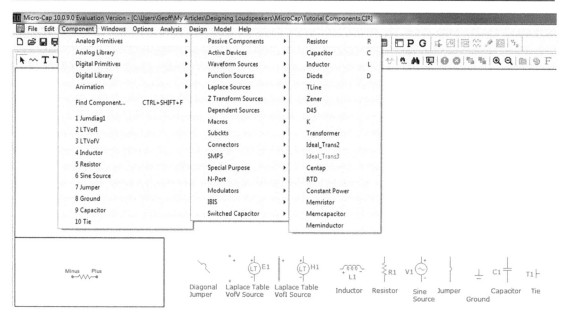

Figure I.1: Micro-Cap Tutorial Components.

I.1   Passive Crossovers

With a passive crossover inductors, capacitors, and resistors are connected in series or parallel combinations between one or more amplifiers and the loudspeaker drive unit(s) to control the final frequency versus amplitude response.

For passive crossover work, we tend to stick with these primitive types as they are always available.

You put them where you want on a circuit layout and literally connect the components as desired. Of special interest are the table components; these are called *Laplace tables*. These can be used to accept a sound pressure level (SPL) versus frequency table or an impedance versus frequency table of actual measurements or simulations of a loudspeaker(s).

These can be found at Component>Analog Primitives>Laplace Sources:

- LTVofV.
- LTVofI.

These have been very useful in simulating loudspeaker crossover circuits, provided the following measurements or simulations of the loudspeakers performance at the desired measurement position(s) are available:

1.   SPL & phase versus frequency.
2.   Impedance & phase versus frequency.

It is necessary to format the measurements or simulations so that they are compatible with Micro-Cap's data format. We find that the best way to do this is with a spreadsheet mapping the data between the two.

The spreadsheet's main output summarises and formats the individual data into combined format that can be pasted into the text page of a Micro-Cap crossover simulation using the LTVofV source (SPL) and the LTVofI source (Impedance), which are connected after the components of the crossover circuits. This can be done manually or when you know exactly what is required; the process could of course be automated.

I.2 Simulating SPL

A simple crossover feeding the LTVofV for simulating the SPL is shown as figure I.2.

We start off with a sine source. Don't worry about the settings. Next, a small resistor to simulate lead resistance, add a small inductor as well if you are concerned with this, then the crossover components themselves, in this case, a 1 mH inductor with a 4u7F shunt capacitor. The loudspeaker's SPL is contained in the LTVoV named 'BassSPL' as a table of values. This is loaded on the 'Text' page and formatted by a spreadsheet, etc.

Clearly in this case, although the SPL response looks okay, the impedance is obviously wrong.

I.3 Simulating Impedance

Next we will look at modelling, just the Impedance, well it looks fine at low frequencies but goes wrong at high frequencies.

This is shown as figure I.3; it is exactly the same circuit as the SPL but it feed's LTVofI to simulate the impedance, however now the SPL is wrong! See figure I.3.

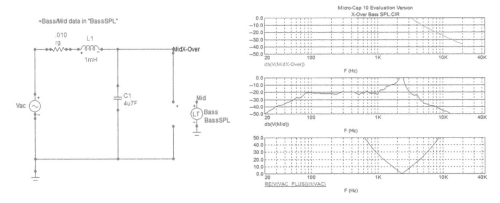

Figure I.2: Micro-Cap SPL Simulation.

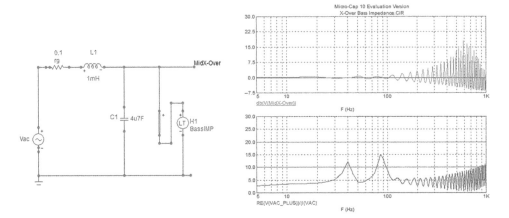

Figure I.3: Micro-Cap Impedance Simulation.

I.4 Simulating SPL and Impedance

Let us combine the voltage and current models.

This is shown as figure I.4.

That looks much better, so we can then produce graphs of the individual sections or drivers in enclosures if we sum or subtract these individual responses of a system from either measurements or simulations using simulated crossovers.

So let us do this now: Click on AC, then Limits, setting each graph up as Y. This is shown as figure I.5.

The individual data coming from either measurements or simulations of each driver or tweeter is saved into the individual sheets. Obviously, this is just a start, but by using a spreadsheet's ability to reformat the measurement or simulated data of individual drivers or tweeters into a form that Micro-Cap can accept, you can build very effective models very quickly and simply.

Then if later you want to model the effective response off-axis at a different distance or on a completely different azimuth, you can run the simulation or measurement(s) and feed the new data in without having to remodel from scratch.

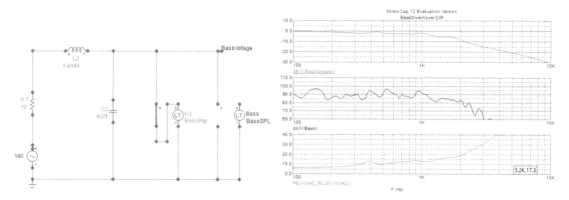

Figure I.4: Micro-Cap SPL and Impedance Simulation.

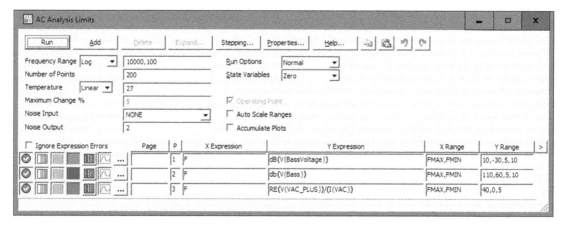

Figure I.5: MicroCap AC Limits.

You can also easily change polarity amongst different 'virtual' loudspeakers to ensure there are no problems or, if there are, catch and sort them out whilst still at the design stage.

Micro-Cap gives you lots of control over the data displayed from the data formats to the number of decimal places.

Note

1. Either analogue or digital/dsp.

PafLS Tutorial

In this tutorial we take a peek behind the curtain that PafLS *very* carefully lays over PAFEC. Now, PAFEC is a very powerful FEA engine, but it is a bit frightening to us at least! As there is just so much that you could do with it—at least in *theory*—but you almost need to know everything before you start! In this appendix, we go through using PafLS to model our subwoofer driver.

PafLS takes a very different approach to loudspeaker simulation; at its heart is the simulation engine behind PAFEC, a fully coupled mechanical to acoustical FEA modelling program.

As we discussed earlier, this means that the acoustical output is coupled to the mechanics and vice versa. However, rather than building a structure from the ground up, it takes a template-based approach for building the underlying model. PafLS thus has a limited range of standard loudspeaker shapes and forms it can solve.

For these forms, it uses various inputs that define the underlying parameters, materials, shapes, and so forth, and maps these into inputs the FEA engine can directly understand. The background PAFEC engine then produces simulated responses, and by comparing these later with measured responses, one can refine a design in detail.

The advantage of this is that some of the typical design choices, cone or diaphragm shapes, suspensions, and surrounds can all be implemented. The downside of this approach is of course that you are limited to models within the templated structure, and these are limited to just a few of the infinite possibilities. What can it do well is axisymmetric, mechanically coupled acoustical simulations of a single driver. It does not do any magnetic modelling, nor can it handle multiple drive units, dual concentrics, or other more unusual situations.

PafLS does not model the voice coil or magnetic structure; rather, these are treated as inputs in the same way that ABEC also can accept the Thiele–Small and electromechanical parameters.

Also, PafLS does not simulate the external shape of, say, an enclosure, which ABEC can; rather, PafLS simulates a loudspeaker driver on an infinite baffle. For the moment, we will just show the results of the LS-WE loudspeaker model.

We will start by opening it from a zero as there are a couple of quirks or problems that need to bypassed before you can use it effectively. File->New opens the New Project dialogue, as shown in figure J.1. Notice the 'Project name'; alongside it there is 'Show existing projects' and 'Location', whilst below is 'Optional project description'.

We will select the LS-WE template as shown in figure J.2.

When we click on the *ellipse* at the right of the location box we can enter a new file location as shown in figure J.3.[1] I would strongly suggest setting up a new folder directly on the desktop as I have shown here; lots of files are generated by PafLS so it's best to keep them together in one folder.

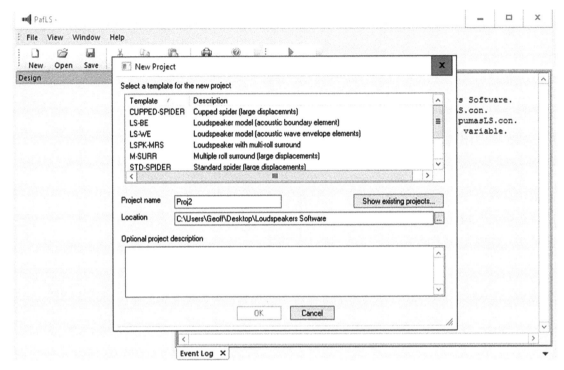

Figure J.1: New Project.

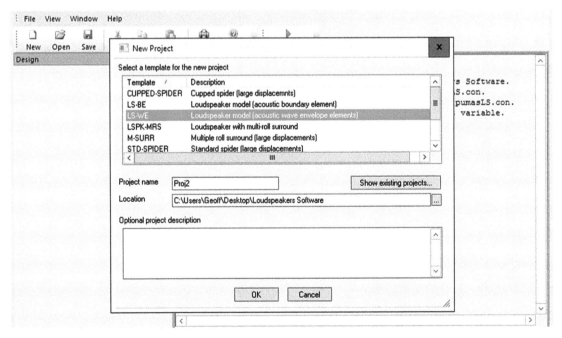

Figure J.2: New Project: LS-WE.

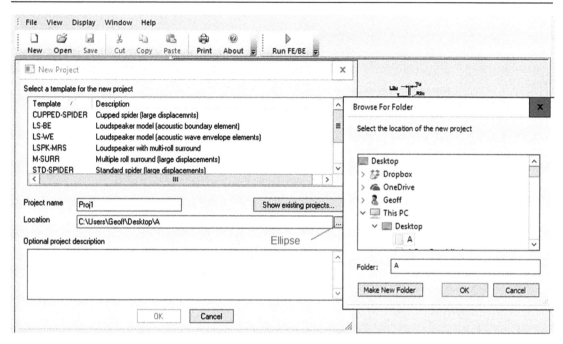

Figure J.3: New Project File Location.

A schematic is shown as figure J.4.

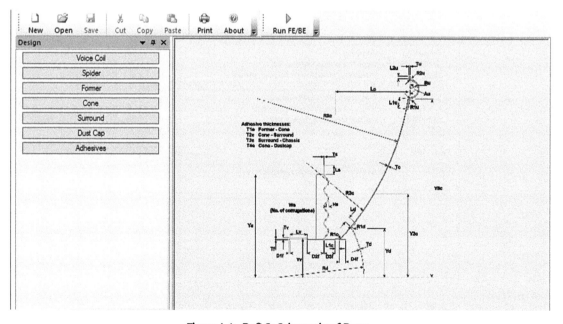

Figure J.4: PafLS: Schematic of Parts.

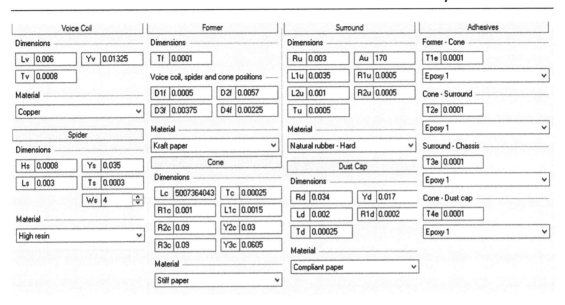

Figure J.5: PafLS: LS-WE Parameters.

The individual parameters that can actually be changed are shown for the LS-WE template accessed by clicking on the individual tabs: Voice Coil, Spider, Former, Cone, Surround, Dust Cap, and Adhesives. These are shown in expanded form for the subcomponents as figure J.5.

As we can see, there are all of the major dimensions that can be changed as well as drop-down selections for materials.

If we click on the Analysis tab (Lower Left), we get to set the frequency points, voltage drive and processors being used as shown in figure J.6.

Click on Sinusoidal Response and we see figure J.7, where we can set start and stop frequencies. We would suggest starting off with small changes first and just one or two frequencies, as it will be quicker and will enable you to check that the model geometry is working before running a detailed analysis later.

The Voltage tab as shown in figure J.8, is where you input not only the voltage at which the simulation will run but also the blocked impedance resistance and inductance. For our subwoofer it is here of course that we would enter the calculated inductance value from FEMM of 28 mH as shown in figure 20.4 on page 87 in Chapter 20 and the force factor *Bl* of 28 Tm as shown in figure 21.11 on page 95 in Chapter 21.

The Parallel Solution tab allows you to change the number of processor cores being used.

After the simulation has successfully run, you can get access to the visual model that represents the tables above, shown as figure J.9. This and the mesh are fully coupled to the input data and physically show the changes entered—very cool!

The predicted SPL is shown as figure J.10 on page 316.

By default, PafLS plots the SPL from 0° to 90°; these can be displayed or not by *checking/unchecking*.

The 0° response simulation is shown as figure J.11.

The acoustic field pressures are shown as figure J.12.

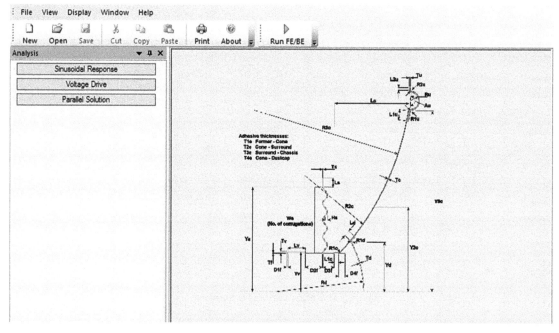

Figure J.6: PafLS Analysis.

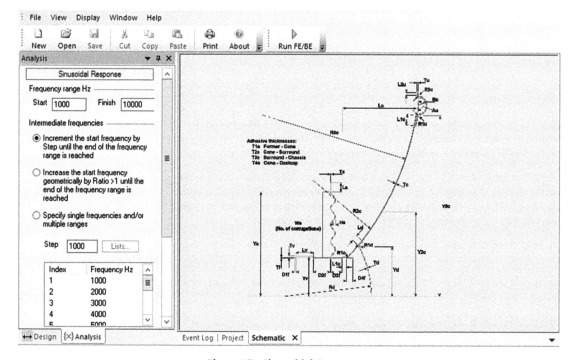

Figure J.7: Sinusoidal Response.

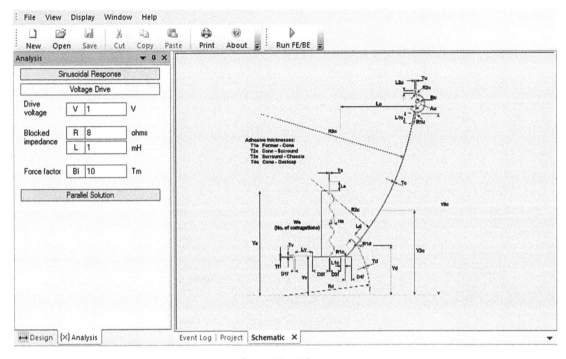

Figure J.8: Voltage.

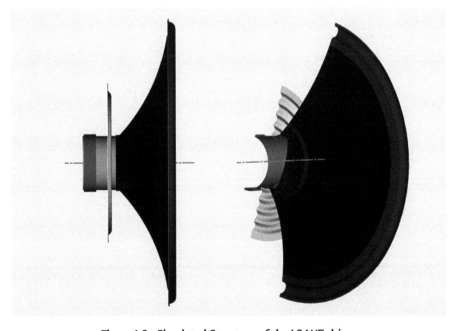

Figure J.9: Simulated Structure of the LS-WE driver.

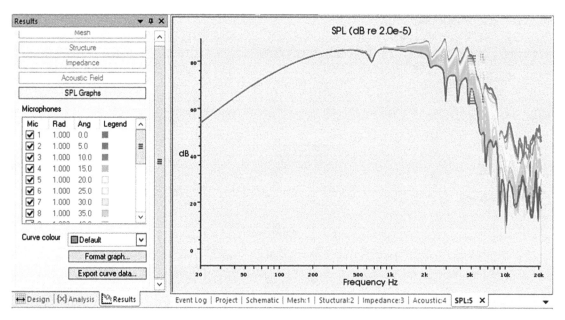

Figure J.10: Simulation of Sound Pressure Level.

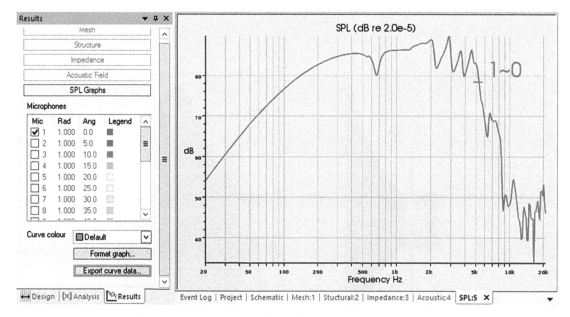

Figure J.11: Zero Degree Simulation of Sound Pressure Level.

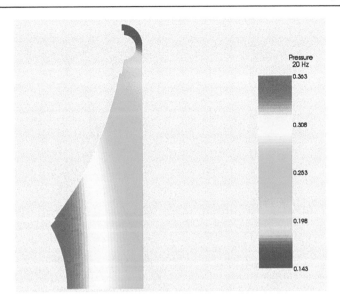

Figure J.12: Simulation of Acoustic Pressure Field.

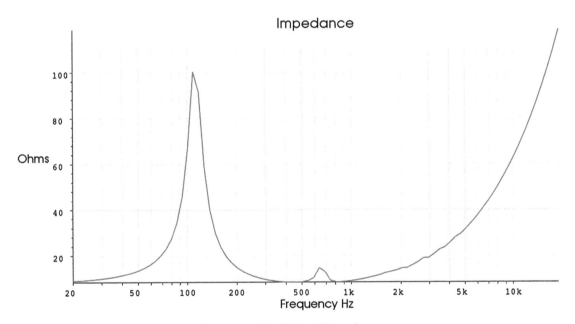

Figure J.13: Simulation of Impedance.

The electrical impedance versus frequency is shown in figure J.13.

The mesh used is shown as figure J.14.

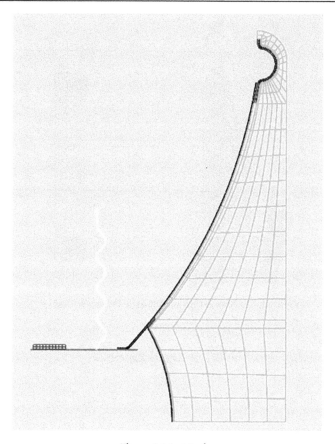

Figure J.14: Mesh.

Note

1. Due to some legacy code with PafLS or PAFEC it is essential to keep to a short file name and path.

Theoretical Bl(x)

In this tutorial, we will concentrate on producing a *Bl*(*x*) simulation. In Chapter 21: Motor Unit on page 88 we skipped over the detail of exactly how this was done, concentrating purely on the results. Here we will examine more closely the details.

We also saw that we could calculate the inductance versus distance or $L_e(x)$ using Lua scripting. David Meeker presents how such a Lua script in the analysis of a motor unit can do this job. However, this sample model and script took over an hour to run and to calculate the *Bl*(*x*) and $L_e(x)$ at just 11 points from -5 mm to $+5$ mm.

Even this took multiple simulations and stitching the results together in a spreadsheet. The same process could be used for *Bl*(*x*); however, this is a very slow process,[1] so as the results need to be displayed as a graph and we need to calculate or design the voice coil anyway, we wrote a spreadsheet to combine the functions of designing the voice coil and calculating the *Bl*(*x*) from the flux distribution of a single FEMM simulation of the motor unit. The problem of *Bl*(*x*) calculation comes down to how all of the active turns collect the magnetic flux available,[2] the core idea behind the spreadsheet being to match the number of points that we run the magnetic simulation for to the voice coil and the magnetic gap we wish to simulate. However, before we get to the detail let us review the overall spreadsheet.

The spreadsheet comprises seven linked individual worksheets. In turn, these are:

- Graphs: where we will see the final results
- Calculations: where we do the calculations
- FEMM ## Point: where we import the forward magnetic simulation
- FEMM Rev ## Point: where we import the reverse magnetic simulation
- Turn Positions: this sheet is redundant
- Copper Wire Gauges
- Voice Coil

We will continue with our design of the subwoofer driver. Although it may seem odd, we find that the best place to start is actually at the voice coil. In Chapter 21 on page 88, we established an initial voice coil design using 0.575 mm copper wire in a 4-layer coil with a wind length of 32 mm using an internal diameter of 75 mm, so let's start to enter this information into our spreadsheet.

We will start with the data *dimensions* that we need to put into the spreadsheet: Magnet depth (mm), top plate thickness (mm), voice coil inside diameter (mm), former thickness (mm), wind length (mm), and number of layers (1 to 4); we enter all of these in the rose-coloured cells. These are shown in figure K.1.

Next we have the wire specifications. Here we can input wire diameter, conductivity, packing density, and insulation thickness, as shown in the tan-coloured cells on figure K.2.

This format allows us to change materials, packing density, and so forth completely. The results of the calculations are shown in the light turquoise cells in figure K.3.

From the first two items, length of gap and number of FEMM points, we now have the core information we need to set up our FEMM model to best effect.

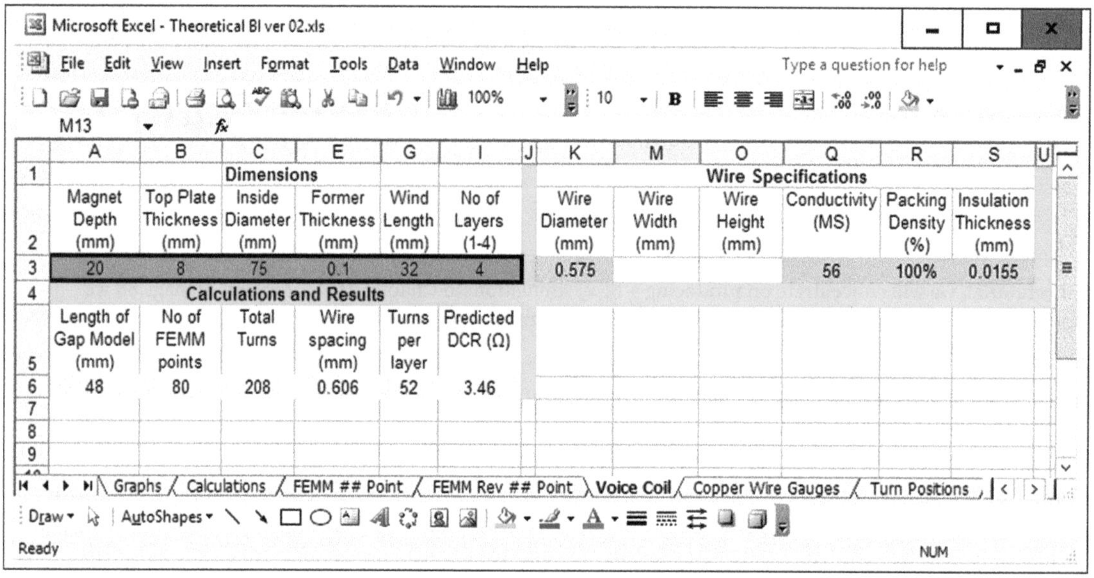

Figure K.1: Theoretical Bl(x): Dimensions.

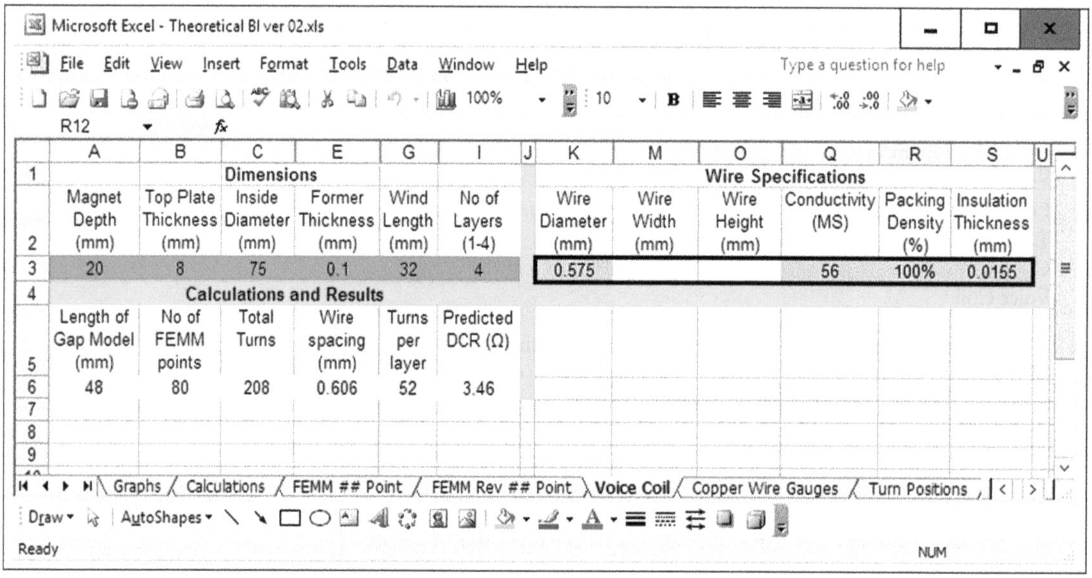

Figure K.2: Theoretical Bl(x): Wire Specification.

So first let us open our original Motor Unit FEMM model answer. We then select the line from +24 mm to −24 mm, which equals the length of the magnetic gap. This is shown as figure K.4.

We then need to click [icon] and enter 208 points and write the data to a text file. This will then match the simulation to the spreadsheet as shown in figure K.5.

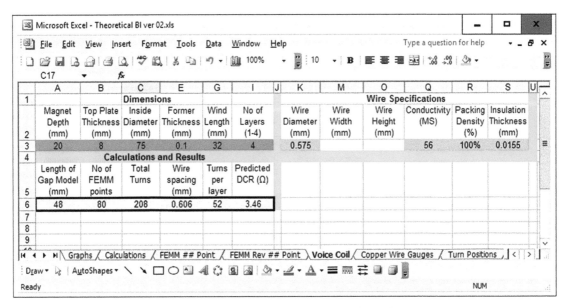

Figure K.3: Theoretical Bl(x): Voice Coil Results.

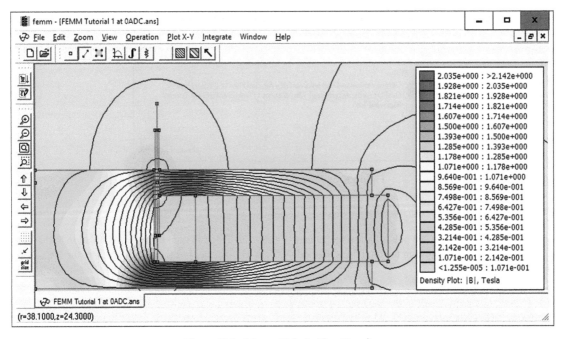

Figure K.4: Motor Unit 1: Flux Density.

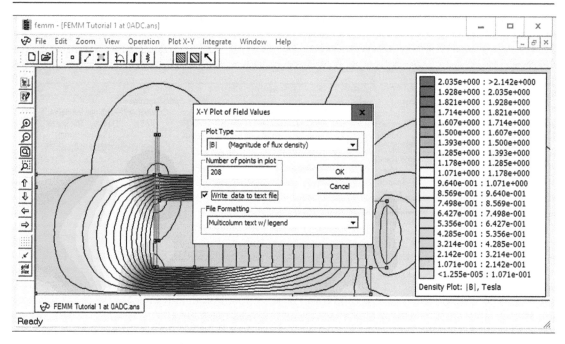

Figure K.5: Motor Unit 1: Flux Density X-Y.

Click OK and enter the name Backward1.txt as shown in figure K.6.

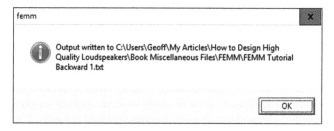

Figure K.6: Motor Unit 1: Backwards.

We then select the line from −24 mm to +24 mm, which equals the length of the magnetic gap. As before, we then need to click ![icon] and enter 208 points and write the data to a text file. This will then match the simulation to the spreadsheet. We then need to sum the total flux collected by the voice coil at all positions. This is where the 208 FEMM points come in.

Click OK and enter the name Forward1.txt as shown in figure K.7.

We now go back to the spreadsheet FEMM ## Point worksheet where we will import the data. Currently there is a bit of adjustment to set up the summation in the calculation worksheet. We open the text files backwards1.txt and forwards1.txt into the FEMM ## Rev Point and FEMM ## Point worksheets; these are linked to the calculation worksheet and the results are displayed in the Graph worksheet. Our initial Subwoofer Motor Unit 1's results look as follows (as shown in figures K.8, K.9 and K.10):

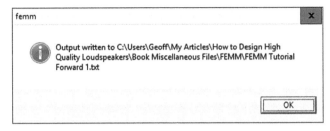

Figure K.7: Motor Unit 1: Forwards.

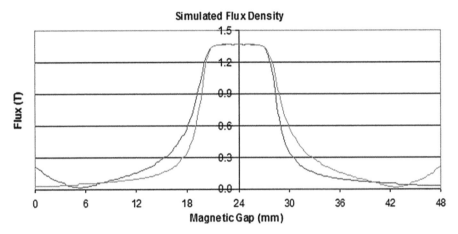

Figure K.8: Motor Unit 1: Flux Line.

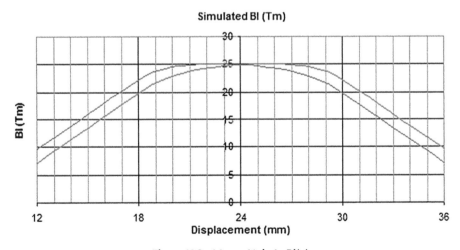

Figure K.9: Motor Unit 1: $Bl(x)$.

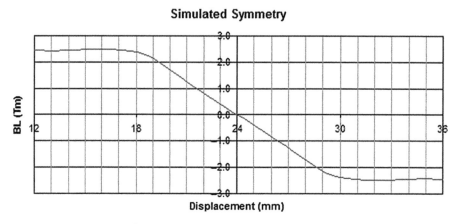

Figure K.10: Motor Unit 1: *Bl* Symmetry.

Notes

1. Our subwoofer motor unit 1 requires around 200 points—clearly impractical.
2. This changes according to the position of the voice coil in the gap, so instead of moving the coil and running multiple simulations, instead we change the selection we sum the flux across.

Statistical Analysis of Loudspeakers

Statistics is all about, well, numbers, so statistical analysis is just using these numbers to tell us more information about a design product or process. So, the first thing is to gather and then record data. Fortunately, pretty well all modern software is capable of collecting the data by default.

In this appendix, we will explore some simple statistical analyses from such data. From the earlier chapter(s) and appendices, you should now be familiar with measuring the sound pressure level versus frequency response or the impedance versus frequency response of a loudspeaker.

The statistical part is nothing more than setting up an analysis on such data from many speakers, and then hopefully you can make some predictions as to whether the loudspeaker(s) you have produced will be of sufficient quality for your use.

There are many approaches to ensuring quality: Kaizen, Continuous Improvement, and Six Sigma being just a few. We like the Six Sigma model. In essence, this takes statistical data from a small batch and calculates what variation this would give when applied to a much larger population. It was originally based upon reliability data.

It is based upon the 'bell curve' or normal distribution whereby for a given tolerance, various percentages will pass: 1 Sigma = 31%, 2 Sigma = 69%, 3 Sigma = 93.3%, 4 Sigma = 99.3%. As can be seen, if we can achieve 3 Sigma, most of the process will be within the desired limits. Strictly speaking, Six Sigma refers to ensuring the statistical stability is less than ±6 standard deviations, corresponding to a single failure rate of 1 in 3,400,000—roughly 3 in 10 million!

As we write this in 2017, we cannot make loudspeakers to ±1 dB over the full frequency range to such standards. However, we have found that ensuring variations are held to ±1 dB and ±3 standard deviations is a challenging but achievable goal for a complete loudspeaker driver.

Therefore, we like to calculate the variability based on ±3 standard deviations. Typically this means a C_p of 1.0, and provided the C_{pk} is also above 1.0, we consider the process is under control. Another useful thing about calculating the ±3 standard deviation values is that simultaneously these can be used to generate the $\pm$ limits. If these are greater than those you would wish, then straight away you have an indication that all is not well with your design or process.

L.1   Probability Curves and Distribution

Before we get into any analysis, let us have a look at a few graphs: Firstly, let us consider a Gaussian distribution from Wikipedia http://en.wikipedia.org/wiki/Gaussian_function. This gives us a probability density curve, as shown in figure L.1 on page 326, with expected value μ and variance σ^2.

These correspond to normalized Gaussian curves with expected value μ and variance α. The corresponding parameters are $a = 1/(\sigma \cdot \sqrt{(2 \cdot \pi)})$, $b = \mu, c = \sigma$.

Here we see a series of typical bell-shaped curves (the x being the standard deviation). We can immediately see that the green curve is offset from all of the other curves whilst the height and the spread of the other varies

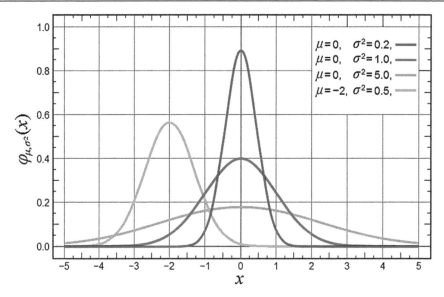

Figure L.1: Statistical Density Curves.

wildly. All of this is fine, but what does it have to do with a loudspeaker? Well, let's look at it this way: If all of the loudspeakers had variations that could be represented by each of these coloured curves, what would that mean?

1. For the blue curve, we could immediately see that all of data would fall within ±1.5 dB of the x value.
2. For the red curve, all of the data would be within ±3 dB of the x value.
3. For the yellow curve, all of the data would be within ±5 dB of the x value.
4. Finally, for the green curve, all of the data falls between −4 dB and 0 dB of the x value. (So very little actually is where it is supposed to be!)

Okay, let's say we have done this. What does all this mean and, most importantly, how can we apply it? Then we have the standard probability curve for a normal distribution. This is shown as figure L.2.

Notice that using the standard probability curve, the first three standard deviations cover 99.73% of a population.

The exact numbers are as follows: $1 * \text{STDEV} = 68.27\%$, $2 * \text{STDEV} = 27.18\%$, and $3 * \text{STDEV} = 4.28\%$. What is this really trying to say to us and what should we be doing with this information?

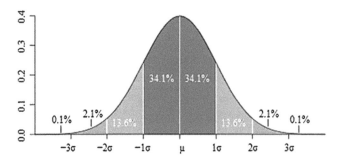

Figure L.2: Statistical Probability Curve for a Normal Distribution.

- Firstly, we should be checking to see that we do have a 'normal' distribution. If we do not, that is a strong indicator that there is a problem.
- Secondly, we should be checking to see that the data is normally centred—that the variation is around the centre or zero x point.
- Thirdly, we should be checking that the spread is not too wide. This, too, would indicate a wide range of variability.

Obviously, the green curve is 'offset' from the nominal value, whilst the yellow curve is much too widely spread. The red curve corresponds to a 'normal probability distribution' whilst the blue curve is very tightly controlled and represents a theoretically ideal scenario.

L.2   Application to Loudspeakers

All of this sounds great in theory, but how can we apply it practice? Loudspeakers do not have a few critical dimensions like a nut or a bolt; instead, loudspeakers cover a wide frequency range from 20 Hz to 50 kHz or more, and there are lots of types of parameters to deal with as well.

So now, instead of simply applying the calculations to a single parameter, let us apply the technique to a much simplified dataset (we have just removed some of the frequencies for clarity in the tables), in this case, sound pressure level versus frequency.

L.3   Seven Bass Drivers

We will take the actual measurement data from a series of bass drivers. We have seven of them, and we think (or have been told) that they are well behaved.

Our specification and most real-life specifications do not say how good they are, so let us find out.

- Minimum value is calculated from MIN(B2:G2)
- Average value is calculated from AVERAGE(B2:G2)
- Maximum value is calculated from MAX(B2:G2)
- Standard Deviation is calculated from STDEV(B2:G2)

where B2 is the first data point or frequency (Hz) through to G2 the last data point at the same frequency (Hz), the process being repeated for all data points and frequencies.

First let's look at it in tabular form, shown as Table L.1 on page 328.

Now let us look at a graph of all seven bass drivers overlaid. This is shown as figure L.3 on page 328. As you can see, it initially looks pretty good.

But let us calculate how good or bad it actually is. We will produce a graph, shown as figure L.4 on page 329, of the minimum, average, and maximum for each data point at the same frequency. A spreadsheet makes it very easy to calculate these results.

It's a bit clearer now. However, it is still difficult to interpret as the data tends to be swamped by the shape of the overall curves, especially if the curve is rapidly going up or down near a data point.

Finally, we can calculate the standard deviation and three times the standard deviation as shown in figure L.5 on page 329. In these curves, we are left with just the variation or standard deviation versus frequency and three times the standard deviation versus frequency.

So now, the underlying shape has disappeared. We can see something really valuable, the variation versus frequency.

Table L.1: Seven Bass Drivers.

Frequency (Hz)	1	2	3	4	5	6	7
19.8	51.5	50.1	43.0	45.4	1.3	43.3	45.2
30.0	51.7	51.8	52.3	53.0	51.3	52.5	53.2
39.6	58.2	56.7	57.6	57.9	56.5	57.5	58.3
49.1	63.3	62.3	62.9	62.9	62.0	62.9	63.3
68.8	70.2	70.0	70.0	69.9	69.4	70.1	70.2
101.1	74.4	74.5	74.6	74.4	74.4	74.6	74.5
197.0	75.5	75.3	75.6	75.5	75.7	75.5	75.4
310.5	76.2	75.9	76.2	76.1	76.2	76.1	76.2
404.3	76.7	76.4	76.7	76.6	76.6	76.7	76.7
692.9	78.3	78.2	77.5	78.4	78.4	77.1	77.2
1043.0	78.5	78.3	78.3	78.5	78.3	78.7	78.7
2075.7	80.8	80.7	80.7	80.4	80.4	80.9	80.8
2933.3	77.6	77.7	77.4	77.3	77.6	77.7	77.7
5107.9	85.8	85.5	87.0	82.4	86.6	86.0	86.2
7229.0	75.6	75.1	74.8	77.0	75.1	75.8	75.8
10234.1	84.1	82.4	84.7	84.8	84.7	84.7	83.2
1190.2	65.9	66.1	66.8	73.9	66.6	69.0	65.2
17854.2	60.6	60.2	61.8	58.9	60.6	63.2	61.6
21999.8	71.6	64.8	67.2	66.8	60.2	71.7	63.2

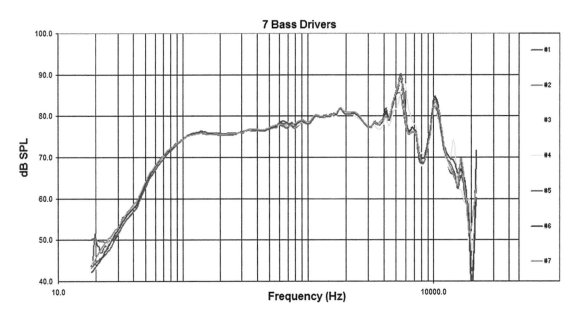

Figure L.3: Seven Bass Drivers.

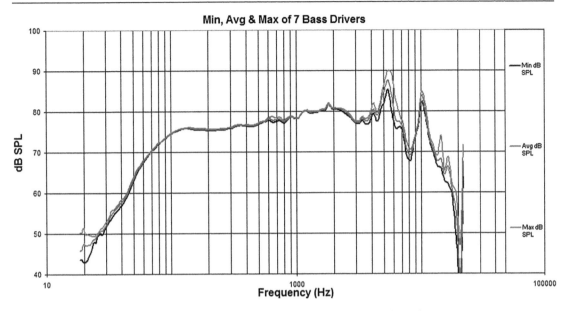

Figure L.4: Seven Bass Drivers Minimum, Average, and Maximum.

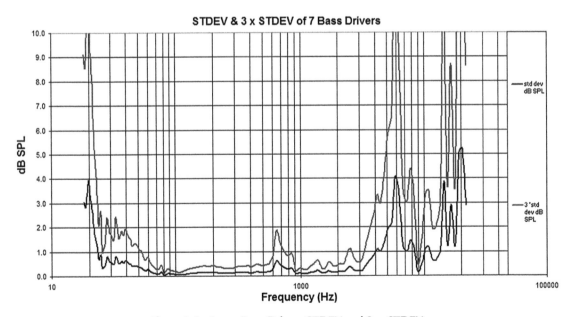

Figure L.5: Seven Bass Drivers STDEV and 3 × STDEV.

From this, we can clearly see that there is a gradual increasing variation below 50 Hz and a bit of a bump between 600 Hz and 900 Hz, indicating some sort of problem, and a more severely increasing variation above 3 kHz.

So firstly, we would ask whether the overall response shape is acceptable?

A real question can now be asked: 'Are these drivers fit for the job they are intended to do?' What does the specification say? ±2 dB from 20 Hz to 2000 Hz. We can now answer equally clearly: 'Yes, they meet this specification.' However, the response is varying excessively above 3000 Hz. The question then becomes: 'Is this driver going be used with a tweeter?'

If it is, where is it being crossed over? If it is being crossed over at or below 3000 Hz, then it's probably okay.

Also, is the tweeter under control where it matters or not? We now have some straightforward questions to ask and answer.

L.4 Fifteen Tweeters

Now we will look at our proposed metal dome tweeter, we had 15 of them. Two however were very low sensitivity, so I will remove them later: No's #3 and #9.

First looking at the statistical output as shown in Table L.2.[1]

Table L.2: Fifteen Tweeters.

Frequency (Hz)	Min dB	Avg dB	Max dB	+3 x STDEV	−3 x STDEV
1008	78.4	82.6	85.1	88.3	76.9
1992	89.1	93.0	94.4	98.7	88.9
3000	87.0	90.8	92.8	96.5	86.3
4008	85.6	89.4	91.3	95.1	85.0
5039	84.2	88.0	89.8	93.7	83.7
7125	82.0	84.8	86.2	90.5	80.9
8484	85.1	87.6	89.0	93.3	84.1
10078	84.1	86.9	88.5	92.6	83.1
13453	85.4	88.2	89.5	93.9	85.1
19031	90.4	95.2	100.1	100.9	87.5
20180	97.0	99.7	102.6	105.4	94.9
21375	94.3	99.6	104.2	105.3	89.4
22641	87.3	95.9	100.5	101.6	85.8
24000	88.4	93.2	96.3	98.9	86.3
25406	88.0	90.8	92.9	96.5	86.1
26930	85.3	88.0	89.5	93.7	84.0
28523	85.6	89.0	91.8	94.7	83.3
30234	83.6	88.9	93.5	94.6	80.5
32016	79.5	87.1	90.5	92.8	79.2
33914	78.4	87.2	90.9	92.8	77.0
35953	64.5	85.6	90.7	91.2	66.9

Let's look at a graph of all 15 tweeters overlaid. This is shown as figure L.6 on page 331. As with the bass drivers, it's a bit of a mess.

So instead let's calculate and graph, as shown as figure L.7 on page 331, the minimum, average, and maximum for each data point at the same frequency. A spreadsheet makes it very easy to calculate the results.

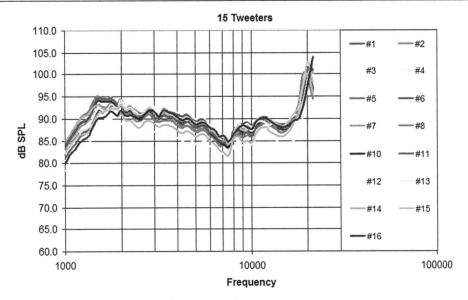

Figure L.6: Fifteen Tweeters.

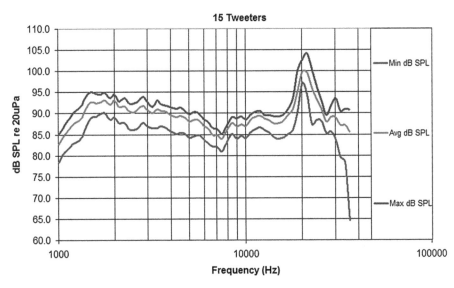

Figure L.7: Fifteen Tweeters Minimum, Average, and Maximum.

Again, it's a bit clearer now and we can see that centring of the average is severely affected by the two low-sensitivity tweeters. So the data is not normally distributed; however, it is still difficult to interpret as the data tends to be swamped by the shape of the overall curves, especially if the curve is rapidly going up or down near a data point.

This now shows us something really valuable, the variation versus frequency. We can begin to see some underlying trends and we can see there is some problem around 1600 Hz.

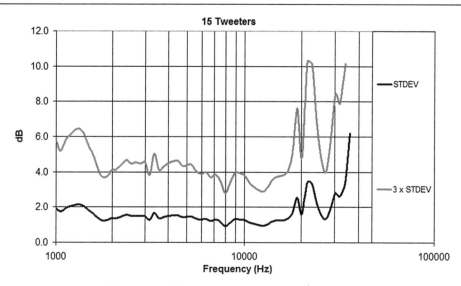

Figure L.8: Fifteen Tweeters STDEV & 3 × STDEV.

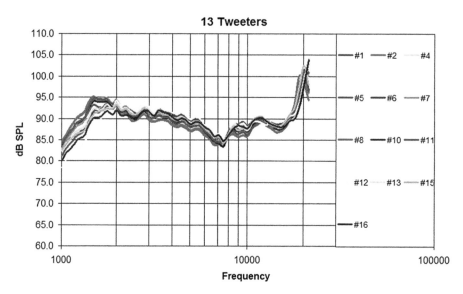

Figure L.9: Thirteen Tweeters.

Finally, we will calculate the standard deviation and 3 times the standard deviation as shown in figure L.8. In these curves, we are left with just the variation (standard deviation) versus frequency and three times the variation (3 * STDEV) with frequency, as suddenly the underlying shape has disappeared.

When we look at the standard deviation, it becomes clear that there is a serious problem with these tweeters around 1600 Hz (resonance damping or ferrofluid?) with over 6 dB variation, but probably worse is over 3 dB variation over the whole frequency range.

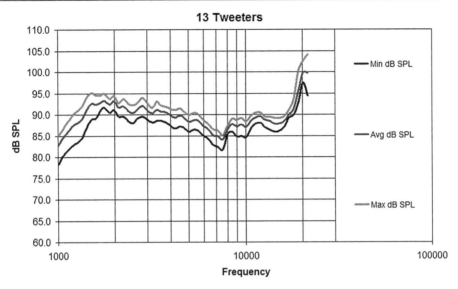

Figure L.10: Thirteen Tweeters: Minimum, Average, and Maximum.

Clearly, this tweeter has some serious problems, which need to be resolved. But before we get too carried away, let us examine the data more carefully and remove the two tweeters that were clearly faulty to see what, if any, effect this has on our analysis.

We will now look at the graph without the two low-sensitivity tweeters, shown in figure L.9.

We can clearly see the overlaid graphs are much more closely aligned, so this would indicate that #3 and #9 were indeed faulty or outliers and *should not* have been included in setting limits.

This is confirmed by the minimum, average, and maximum curves. The average is closer to the centre but it's clearly not completely centred. However, there appear to be quite a few samples close to the minimum, so it would not be safe to exclude them. It would be safe to say that there is a significant variation in sensitivity with this tweeter as shown in figure L.10.

This is confirmed when we look at the standard deviation and three times standard deviation curves—effectively these predict that this tweeter *will* have a sensitivity variation of greater than 3 dB from 1000 Hz to 10 kHz!

L.5   Summary

In this case, clearly the bass driver is pretty good up to 3 kHz but the tweeter could *not* be considered remotely ready for use as shown by figure L.11.

So, we can see the real power of this analysis is in being able to predict the behaviour of more 'normal' data—doing so with relatively few samples and doing so for a much larger 'population'.

The analysis presented here is simplified, as our aim here was to explain the underlying principles.

A proper analysis would ideally look at all of the following:

- SPL versus Frequency
- IMP versus Frequency
- Small Signal Parameters

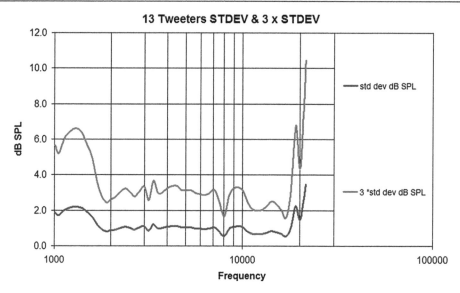

Figure L.11: Thirteen Tweeters: STDEV and 3 × STDEV.

- Large Signal Parameters
- Rub and Buzz
- Distortion
- Furthermore, by relating these parameters against those of 'reference' units, we can calculate:
  - C_p
  - C_{pk}

This process has been extensively used to analyse data amongst many suppliers of drivers, tweeters, and cabinets for many years and has been of proven benefit in tracking down and resolving problems.

Note

1. Note this is *Real Measured Data*.

WinISD Tutorial

In this tutorial, we go through the steps required to enter a loudspeaker's parameters into the program and then use it to simulate the amplitude, SPL, impedance, and many other response curves. It is also useful for its database of loudspeakers. It is primarily designed for subwoofer box and driver as well as simple crossover work.

WinISD will also give an indication of high-frequency performance through its use of additional filters to model high-frequency roll-off. It can be downloaded direct from: www.linearteam.dk/default.aspx?pageid=WinISD. See figure M.1.

From our point of view, its most significant capability is that you can build your personal database of virtual or real drivers based purely upon the Thiele–Small parameters.

However, it can be a bit difficult to do this unless you let WinISD calculate some of the parameters you already 'know'. An example would be if you have values for Q_{es}, Q_{ms} and Q_{ts}; please only enter two of these and let WinISD calculate the third one.

Or for F_s, M_{ms} and C_{ms}, also enter either S_d or diameter. Doing it this way will save you a lot of trouble. We will start off by creating a new project. This is shown as figure M.2 on page 336.

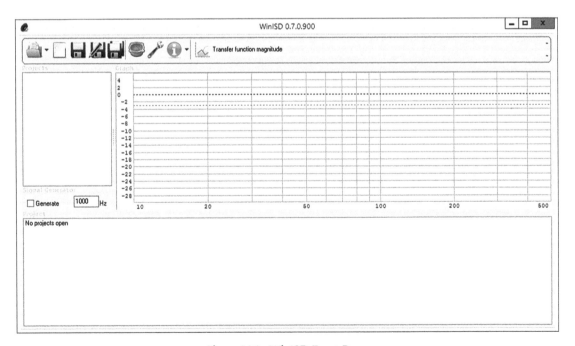

Figure M.1: WinISD Front Page.

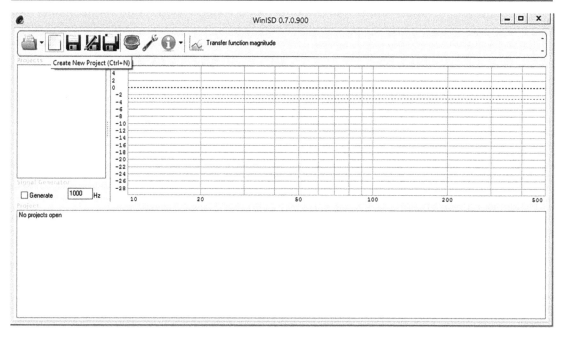

Figure M.2: Create New Project.

This brings up the Select Driver dialogue. This is shown as figure M.3 on page 337.

We will click Add New and see the driver editor. This is shown as figure M.4 on page 338.

Here you can enter essential information about the particular loudspeaker you wish to model. This is especially important if you are working on an early design as there may be many variations and it is important to keep track of them. Please use the comment field, otherwise you can easily end up with orphan data—a curve for which you have no idea what it relates to.

The Dimensions tab shown as figure M.5 on page 338 is very useful here. We would recommend that you fill it in at the beginning and then at each change, as then it will be up to date as your model progresses.

The Advanced Parameters tab shown as figure M.6 on page 339 is mainly about thermal and dynamic modelling—but remember this is a lumped parameter model so it shows its limitations here.

Mainly, though, you will probably concentrate on the Parameters tab shown as figure M.7 on page 339.

This is very logically laid out into four main sections:

- Thiele/Small parameters
- Electromechanical parameters
- Large-signal parameters
- Miscellaneous parameters

You can start from either the Thiele/Small or the electromechanical parameters. As you enter a value, WinISD then calculates the other appropriate values in other fields to match your entered value. Please *do not* attempt to enter into WinISD a full series, say Q_{es}, Q_{ms}, and Q_{ts}—your values will probably not be what WinISD expects. We have found it is best to enter the minimum number of figures you know and then to let WinISD get on with the rest.

Figure M.3: Select Driver.

Let us continue with our subwoofer example: What do we know at present?

- $Bl = 28$ Tm
- $X_{max} = \pm 12$ mm
- DCR $= 3.46$ ohms
- 4 layers
- $F_s = 38$ Hz
- $A = 950$ cm
- Closed box—so $Q_{ts} = 0.7071$
- $M_{ms} = 0.2944$ kg
- $C_{ms} = 0.050598E - 3$ mm/N

Are these enough? Well, let us see: (1) Let's enter Q_{ts}, (2) DCR $= 3.46$ ohms, (3) $C_{ms} = 0.050598$ mm/N, (4) $M_{ms} = 294.4$ g, (5) $Bl = 28$ Tm, (6) $V_{as} = 64$ l, and lastly (7) estimate $Q_{ms} = 5$.

Notice—I am *not* entering the resonance frequency of 38 Hz; we will *let* WinISD work that out. It will also give us a check to ensure that the calculations are approximately correct.

In figure M.8 on page 340, we see these in green and those calculated by WinISD in blue.

When we enter 64 litres as the closed box volume, the SPL versus frequency is plotted as shown in figure M.9 on page 340.

Figure M.4: WinISD Driver Editor.

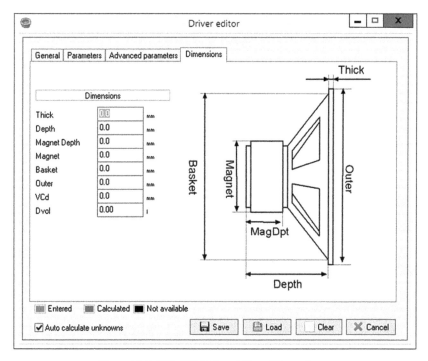

Figure M.5: WinISD Driver Editor Dimensions.

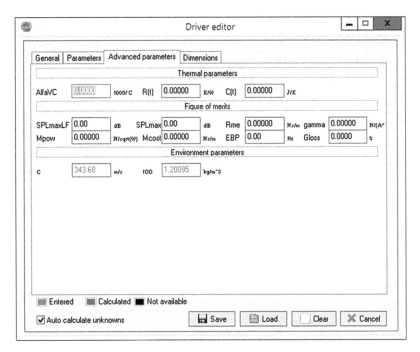

Figure M.6: WinISD Driver Editor Advanced Parameters.

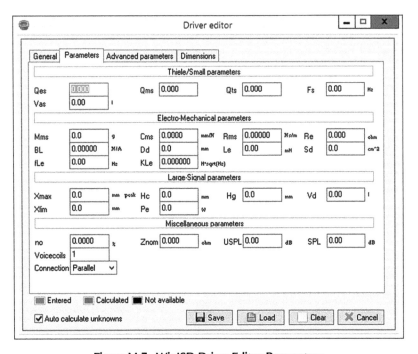

Figure M.7: WinISD Driver Editor Parameters.

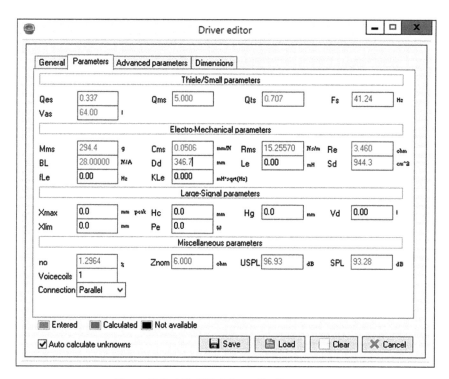

Figure M.8: WinISD Subwoofer Parameters.

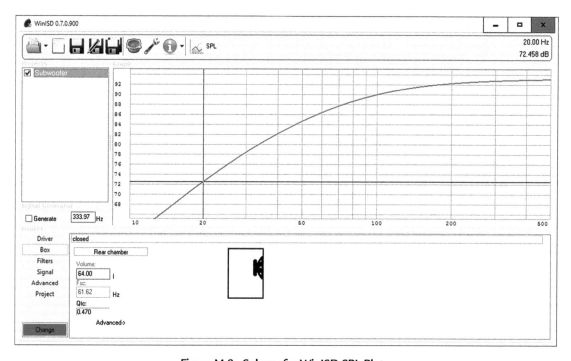

Figure M.9: Subwoofer WinISD SPL Plot.

Index

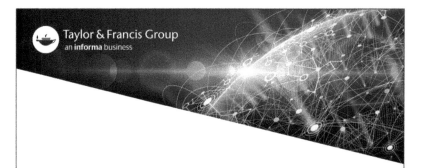

Taylor & Francis eBooks

www.taylorfrancis.com

A single destination for eBooks from Taylor & Francis
with increased functionality and an improved user
experience to meet the needs of our customers.

90,000+ eBooks of award-winning academic content in
Humanities, Social Science, Science, Technology, Engineering,
and Medical written by a global network of editors and authors.

TAYLOR & FRANCIS EBOOKS OFFERS:

A streamlined
experience for
our library
customers

A single point
of discovery
for all of our
eBook content

Improved
search and
discovery of
content at both
book and
chapter level

REQUEST A FREE TRIAL
support@taylorfrancis.com

 Routledge
Taylor & Francis Group

 CRC Press
Taylor & Francis Group

Milton Keynes UK
Ingram Content Group UK Ltd.
UKHW052029141024
449569UK00017B/750